THE SINGLE HELIX

Also by Steve Jones

Genetics for Beginners
(with Borin van Loon)

The Cambridge Encyclopedia of Human Evolution
(with Robert Martin and David Pilbeam)

The Language of the Genes

Genetics: A Third Level Course
(with Kay Taylor)

From Genes to Cells
(with S. Bolsover, J. Hyams and E. Sheppard)

Almost Like a Whale

Y: The Descent of Men

THE SINGLE HELIX

A Turn Around the World of Science

STEVE JONES

LITTLE, BROWN

LITTLE BROWN

First published in Great Britain in November 2005 by Little, Brown
Reprinted 2005

A CIP catalogue record for this book
is available from the British Library.

ISBN 0 316 73193 5

Typeset in Baskerville by M Rules
Printed and bound in Great Britain
by Clays Ltd, St Ives plc

Little, Brown
An imprint of
Time Warner Book Group UK
Brettenham House
Lancaster Place
London WC2E 7EN

www.twbg.co.uk

Contents

Preface: The Single Helix

The title of this book is half-borrowed (or half-stolen) from the second most famous work in the whole of popular science writing, which means at least that it is not a brief history of anything. James Watson's *The Double Helix*, published in 1968, is a racy account of the race to discover the structure of DNA. It showed for the first time that science can be fun and that at least some scientists are human. *The Single Helix* pays homage both to Watson's title and to his theme, and if my book is half as good as its original I shall be well pleased.

Homage is a dish best eaten cold, and I tried to use a helical title in my very first piece of writing on popular science, twenty-five years ago. The journal *Nature* – doyenne of scientific journals and, like her sisters, rather too expensive and rather too carefully made up – has a 'News and Views' section that interprets research for the non-specialist. In 1980 I ran a conference on the genetics of snails and *Nature* asked me to write an article on the event. I produced a suitably leaden piece and – in a fatal attack of flippancy – called it 'The Single Helix', after the Latin name of a snail common in British gardens. At least, I thought, I would get more reprint requests than for any of my real scientific papers and it might even amuse James Watson. *Nature* disagreed: the idea was too frivolous for its august pages, and the piece appeared under a duller heading.

That 'News and Views' article was the first of many – perhaps too many – that I have written on evolution and on genetics. It marked the end of one career and the start of another. I continued to add further bricks to the tottering edifice of molluscan knowledge but with the advance of molecular biology felt my research to be more and more on the margins. To make things worse, money became harder to get and I grew tired of submitting grant applications that ended up, one after the other, in some bureaucrat's bin.

If there is one rule of malacology, the study of molluscs (from the Greek *malakos*, soft and floppy, a phrase much used in Greek slang), it is that nobody gets anywhere until he gives it up. Edgar Allan Poe's *The Conchologist's First Book* outsold his early horror stories, while Lewis Carroll was an amateur collector (his bandersnatch, frumious as it may be, is oddly close to the German *Bänderschneck,* for 'banded snail'). Robert Louis Stevenson claimed that 'It is a more fortunate destiny to have a taste for collecting shells than to be born a millionaire' and Jean Piaget, the child psychologist, was an expert on the freshwater snails of Switzerland (the effects of environment on shell shape fed into his notions about infant learning). In more recent times, Stephen Jay Gould spent years with such creatures before ascending his Olympus of orotundity; and I myself corresponded with Alex Comfort about shell pigments while he was, without my realising it, writing that singularly unmalacological volume *The Joy of Sex*.

Plenty of middle-aged scientists face such crises of confidence about research. They face two choices, administration or vulgarisation. Vast though the opportunities in the first have become, I turned to the second.

Science is a broad church full of narrow minds, trained to

know ever more about ever less. We write about what we understand. Any technical paper anchors itself within a framework of new data and evidence from the literature. A letter in *Nature* may appear opaque to most lay-people, but it is equally obscure to most scientists (which is why the 'News and Views' section was invented).

Even so, for those in the know the most technical publication, written in the clumsiest prose, can be read as a work of grace and elegance. This book, though, is an attempt to escape from the arcane language of research and to translate it into plain English; to take the ideas of science into the wider world of words. Its one hundred easy pieces are variations on the theme of modern science from astrophysics to anthropology and genetics to geochemistry. Almost all attempt to explore a field about which I myself knew (almost) nothing before writing an essay upon it (although I have allowed a few snails to creep in here and there). Most are based on edited and updated versions of my 'View from the Lab' column, which has appeared on the science page of the *Daily Telegraph* for more than a decade. I thank Roger Highfield, the newspaper's science editor, for his editorial advice over the years.

On looking back I am depressed to see how many of those columns were devoted to carping, the age-old prerogative of the academic, with complaints about cash mixed with generalised ill will. Such grumbles soon become dull and none has made it to this book. Instead I concentrate on science itself. Like politics, sport and gossip, it never stops – which is a great help for a writer. Unlike them, the subject has one huge additional advantage. *The Single Helix* follows the twists and turns of science as it moves in the only direction it understands: forwards.

THE SINGLE HELIX

On Darwin Airlines

I have had for many years, albeit without formal training in the subject, a deep interest in the design of large passenger aircraft. Sad to say, because of the blind prejudice of the aviation establishment not one of my prototypes has yet been built. Without giving too much away, I know, as a biologist, that feathers are the best way into the air; and, provided with enough cash by a rich bird fancier, I would be happy to educate a new generation of engineers to believe in my ideas – or, in all fairness, at least to afford them the same weight as the now-controversial Boeing Theory.

And that, in essence, is the argument of the creationists who have hijacked a state-funded English school. The staff of Emmanuel City Technology College (a 'Beacon School') in Gateshead believe that both evolution and creationism – a world billions versus one thousands of years old – are 'faith positions' that should be presented to children as theories of equal value.

Their claim is, needless to say, garbage. The whole of science unites to prove them wrong. To teach biology without a belief in evolution is like giving a course in aeronautical engineering with certain doubts about the force of gravity or in English with no confidence in grammar. Still, there they are, and British youngsters are as a result being fed subsidised lies about science.

Such drab proponents of stupidity might be dismissed as clowns or cranks but they are dangerous. To biologists, of course, who expect their students to know the basic facts of their subject when they come to university (is Emmanuel really an official educational Beacon here; and should it be allowed to remain one?), but also in a wider sense.

Creationists have an argument not with science but with argument. I often go to conferences about evolution. Like all scientific meetings, they are filled with disagreement. One of the most bitter is about Darwin's 'mystery of mysteries', the nature of species. I don't know the solution, and I am not sure that anyone else does, but we keep talking and doing experiments and, with luck, some day it will emerge.

Compare that with the dead-eyed certainty of the biblical literalists. There is no room for discussion, because the answer is written in a big book. Of course, to keep the powers-that-be at bay they have to admit that there is another theory out there, but as Christians they just *know* it is wrong. A hundred and fifty million Americans go along with them (as I said to my American publisher – who was not amused – I don't mind if they burn my own Darwinian books as long as they buy them first). My academic colleagues across the Atlantic now have to waste large parts of their first-year courses trying to put right the damage done by the educational Taleban who rule not just in a single city but across much of the nation.

I claim no expertise in theology (but – hey – these guys don't hesitate to lecture about evolution) but to me such people do more harm to religion than to science. What if a bright young child realises later in life – as he will, if he continues to study biology – that he has been fed untruths by his teachers? Why should he believe anything else that the Good Book says? Why, indeed, remain a Christian at all?

The Bishop of Oxford and many of his colleagues see that point and have spoken eloquently against the fundamentalists; but, asked about school creationism, Mr Blair, the Prime Minister who lit the beacon for such outstanding schools, could do no better than a few dismal weasel-words: 'I think it would be very unfortunate if concerns over that were seen to remove the very, very strong incentive to make sure we get as diverse a school system as we properly can.'

I once saw a bumper sticker that said, 'If you think education is expensive, try ignorance.' Now we know the price of ignorance in the PM's mind: two million pounds – the donation by a Christian businessman that set Emmanuel on its primitivist path. That seems a pretty good bargain, at a couple of grand a kid. Even better, for that investment you get a new crop of innocent minds to poison each year – and, with plans to set up yet more schools with the same system of belief, stupidity will soon be available wholesale.

Emmanuel College's grubby secret came out when a creationist conference was held there. The programme of the next meeting, in Pittsburgh, summarises the nature of their case: 'Papers dealing with the age of the Earth must be from a young-Earth perspective. Papers from an old-Earth perspective will not be considered.' And, as the Pennsylvania-bound Jones 666 (I believe in biblical numbers for my aircraft) flaps into Newcastle Airport with its plumes waving, I look forward to ushering the Gateshead delegates on board.

O Happy Day

A year or so ago I celebrated National Marriage Week in the traditional way, by getting married. It was a perfect excuse; I managed to get out of two tedious academic meetings by pleading a long-standing engagement (about twenty-five years in this case).

The actual event happened on a Friday, the thirteenth of February, in the romantic milieu of Camden Town Hall. That date was the only one available. Saturday was, as might be expected, booked up because it was the feast day of Valentine, the patron saint of bee-keepers, plague, epilepsy and the betrothed, and the other days were pretty busy, too. Friday the thirteenth was a void.

But why? The Italians omit the number thirteen from their National Lottery, but can the bluff populace of England be so superstitious? Its malign reputation is said to go back to the Pharaohs (thirteen lunar months in a year), or perhaps the Vikings (the God of Mischief caused mayhem when he gate-crashed a dinner for twelve) or the number of seats at the Last Supper. Whatever the legend's origin (and the notion of especially bad luck on a Friday when it falls on the thirteenth is but a century old), plenty of people seem to believe it.

Many have claimed that the day is indeed bad for your health. A Finnish doctor discovered that, from 1971 to 1997, the rate of fatal traffic accidents went up on Fridays bearing the evil date compared with the thousand and more not so cursed. Oddly enough, the effect seemed to apply only to women, for Finnish men were immune to the baleful vibrations of the malign day. As the author says: 'It is not inconceivable that on Friday the thirteenth women who are susceptible to superstitions obsess that something unfortunate is going to happen, which causes anxiety and the subsequent degradation of mental and motor functioning.'

The fate of the Finns is based on statistics, so the story must be true; but it faces a problem. A Laura Buxton (aged ten) once released a balloon that was picked up by another ten-year-old Laura Buxton a hundred miles away. Amazing – but millions of balloons escape and most of them have a more ordinary (and less public) fate. The fuss about Laura emerges from the statistical fact that unlikely events are bound to happen now and again, and that as more information is gathered they have a greater chance of being noticed.

The Fatal Friday problem comes from multiple testing; from picking at the data in finer and finer detail until something turns up. A scientist might work on the association of blood pressure, gender, smoking, race, exercise, diet, height, weight and day of the week. He gathers the facts, fires up his computer and, to his amazement, finds that, with a chance of error of just one in twenty (the level accepted for most scientific papers), blood pressure goes up on Tuesdays. A quick letter to *Nature* and fame is around the corner.

Alas, the manuscript is rejected, for it has fallen into the Laura Buxton trap. Because so many variables were examined, the danger of a false positive increases (by quite how much is the subject of arcane statistical arguments). Finnish women may have done worse on unlucky Fridays, but in Britain, with its mass of information on car accidents, one might just as well find that red-bearded Scots dwarves were more likely to be killed in leap years – a fact that means nothing.

Early in my career I studied (it seemed a good idea at the time) the fit between snail genes and microclimate in Romania. With immense effort I gathered information on sun, rain and the rest, and to my delight found a fit between the frequency of one shell type and the incidence of thunderclaps. Alas, I was laughed out of court at a conference and the result never made it into print. In the same way, as the double helix was charted, disease genes were confidently ascribed to particular chromosomes because they seemed to be associated with one of the thousands of mapped DNA variants. In many cases the fit was quite accidental. It evaporated, along with its discoverer's reputation, when more families were studied.

Now, technology has made matters worse. Genes can be seen at work. Some seem to be active mainly in tumours, or after drug treatment – but how many of the fits are false discoveries due only to multiple testing of the thousands of lengths of DNA involved as they switch on and off? The web is full of statistical packages to help, but they are loaded with health notices. Mere biologists need a statistical counsellor to see them through the maze. Too much for me: I'll stick to the snails.

And what of the lucky thirteenth? Nothing to worry about on the crucial Friday, I'm glad to report; fortunately, my wife (as I have learned to call her) cannot drive and I do not own a car, so we went to the ceremony by bus. Fingers crossed for the next twenty-five years!

Seeing with Bullets

The interrupted Hundred Years War that marked the twentieth century was ignited in 1904 by a short but bloody struggle between the Russians and the Japanese. It began with the destruction of the Russian fleet in Port Arthur, on the southern coast of China, thirty-seven years before Pearl Harbor. The conflict marked one of the century's first scientific breakthroughs.

Wars have always been good for the medical profession, for politicians see the value of health and unusual patients appear in the surgery. Research on the immune system got a huge boost from attempts to transplant skin to burned soldiers in the Second World War. The research tool in 1904 was the new Russian rifle, the Mosin-Nagant, the first to fire small, fast and lethal rounds.

Such bullets could enter an enemy's skull without shattering it. Instead, they caused local damage. A Japanese doctor used patients presenting with Mosin-Nagant syndrome to show, for the first time, that damage to part of the brain near the back of the head leads to blindness.

Real progress had to wait for the First World War itself. A chart of the skull's interior made using high-velocity bullets as surveying instruments showed how each point in the eye maps to a separate part of the brain. An atlas with a missing page (as anyone with a battered $A–Z$ soon notices) is a vexatious thing. In the same way, a damaged brain map leads to a loss of awareness about part of the outside world.

Since Freud, everyone has heard of the unconscious mind, even if they mock the idea. Now we know that it has a life of its own. In 1904 the Hull fishing fleet was shelled in the North Sea by a Russian admiral on his way to Japan who, for no obvious reason, mistook drifters for enemy torpedo boats. His fear of attack overwhelmed his common sense and his mind rejected the evidence of his eyes.

Mind and eye are intimately linked and in some ways the eye is no more than a forward projection of the brain. When the tie between them breaks down, the result can be disastrous. For most people, blindness is a simple and sinister thing – the complete loss of a conscious experience; but the unconscious creeps in here, too.

The opening of the Cambridge Psychological Laboratory was described as 'putting the soul on a pair of scales' and the first attempts to measure sensations caused a scandal. Now consciousness itself has been put in the balance. A light is switched repeatedly on and off, and at the same time slowly dimmed until the subject says that he cannot see it at all – but, when asked to guess whether it is on, more often than not he gets it right! The unconscious has perceived something not sensed by the conscious mind.

The fate of soldiers with a bullet in the brain helps us to understand how it does the job, for such unfortunates may share that strange ability. The damage has destroyed the million or so nerve fibres which pass the message from the outside world to the visual part of the brain, but has left unscathed a secondary set of nerves from the eye that go to other places inside the skull. They continue to transmit information, even if those who receive it do not refer to it as 'light'.

Such unseeing vision is called 'blindsight'. Many people blind as a result of brain damage do much better than would be expected by chance when asked to say whether a

light is switched on or not, even though they have no conscious idea that they can see. Some can even tell whether an
object in their visual field is moving and what size or colour
it may be.

For a blind person to identify the colour of an orange
seems miraculous, but it is not. Their brain has enough
information to make decisions about light, but not enough
to 'see'. Normal people playing the psychologist's game with
dim lights need stronger evidence – a brighter light – before
the information invades their consciousness, but even a stim-
ulus well below the crucial limit is in some way perceived.

Is seeing beyond the edge of consciousness seeing at all?
Is blindsight just a damaged kind of sight, or is it something different? And – as the Mosin-Nagant bullet was the
first to hint – is just the back of the brain needed to perceive
the outside world, or does the whole brain have a stake
in vision? All this is part of the new science of 'neuro-
philosophy', which explores the once unknowable.

Francis Crick (the DNA man) spent twenty years on such
questions, and wrote a book on his research subtitled – with
a certain immodesty – *The Scientific Search for the Soul.*
Although he did not find that mythic organ, he thought
that the soul is nothing but blindsight writ large; no more
than a collection of insensible perceptions which can be
weighed in a scientist's scales.

Some of the sailors at Port Arthur may furtively have
read Freud's new book on the interpretation of dreams, in
which such experiences are seen as a sexual message from
the unconscious. Many were, thanks to the new Russian
rifle, doomed to take part in the first experiment to show
that the brain does indeed have hidden depths, even if the
soul is not hidden within them.

Heavyweight Science

Two centuries ago Henry Cavendish weighed the Earth and everything upon it: or, to put his work in more scientific terms, he measured the gravitational constant, G.

Gravity (Newton's 'most subtle spirit') is the glue that makes the cosmos stick together. When apples fall and the Earth spins round the Sun they follow a simple rule. Each object (Earth and apple, or Sun and Earth) attracts the other with a force equal to their mass multiplied together, over the square of the distance between them, with the whole lot multiplied by Big G, the famous constant itself. That discovery was the first step towards the unification of physics: the integration of what seem very different things into a single framework. The Grand Unification which might join all the forces of the universe in this way is not yet with us, but – if it exists – Newton took the first step towards it.

Cavendish came up with a clever experiment. He took two small metal balls connected by a rod to make a dumb-bell and balanced it from a thin thread. A hefty lump of lead was then placed on each side. Their gravitational attraction made the dumb-bell twist. As the resistance to twisting by the thread, the distance between the bodies and the mass of each one are all known, the extent to which the test object rotates can be used to work out the constant G.

Simple enough, and one assumes – given its importance to physics – the value of that crucial figure must by now be fixed to the twentieth decimal place. In fact it is not. G is by far the least well measured of all physical factors. If the size of the Earth were as badly known as is its weight (which depends on the constant) we would be able to measure it to an accuracy of within only a mile – compared with the precision of a single inch in our estimates even of the far larger distance from Earth to Moon.

The second centenary of the Cavendish experiment in 1998 marked the start of a new international effort to sort out the real value of G. Although the disagreements among those who measure it continue, the elephantine truth may soon lumber into view.

Odd as it might seem to anyone who has tried to lift a sack of cement, everyone agrees that the notorious G is very, very small, the weakest of the physical forces of the universe, and is hard to measure. Many more or less eccentric methods have been tried. One looks rather like the Cavendish apparatus, although the dumb-bell is made to swing back and forth as it dangles from a fine quartz fibre and the effect on its rate of oscillation of nearby weights made of tungsten measured. The Germans have done a similar experiment with a test body afloat on a bath of mercury. Another trick is to hang a couple of pendulums next to each other, wheel in a massive object on each side and see how far the hanging balls move apart.

The latest approach involves a small steel ball with a reflector attached, off which a laser beam is bounced. The ball is dropped thousands of times through a hole in a half-ton tungsten tube. When the massive tube is below the falling sphere, the ball falls faster; and when the ball is below the lump, the gravitational attraction of the great tungsten mass slows it down. Comparing the rate of fall above and below the metal block is a measure of G. The experiment was done in Colorado, in a vacuum, in a place far from earthquakes or other disturbance. The machinery was sensitive enough to pick up the effects of the ocean tides, a thousand miles away, and even the mass of the air above as the weather changed.

Once estimates of Big G varied a lot. Now almost everyone agrees; its value (in its own units of mass, length and time) is 6.67 times ten to the minus eleven, which is, it is hard to deny, pretty small. More important, the figure stays pretty much the same over very different scales. A modern 'torsion pendulum', as the Cavendish apparatus is called, can measure gravity over distances of a tenth of a millimetre, while other experiments bounce lasers off reflectors placed on the Moon to measure its precise distance from the Earth in relation to its mass. Both produce, within the limits of experimental error, the same value of G (which annoys proponents of abstruse ideas such as string theory, who feel that gravity should bend to their ideas).

Even so, the Germans in their mercury bath find the figure to be about half a per cent more than anyone else does. The difference seems trivial, but if they are right the mass of the Earth would increase by ten million billion kilograms. Whether Germans naturally find life more ponderous, or whether it all turns on a mistake in the sums, nobody yet knows.

Nine Lives – for Some

I am, every morning, impressed and enraged by the genetical exhibition that urinates on my plants. The DNA that takes part in this repellent pastime is contained within cats. Almost every day yields a tabby or two, some black-and-white ones, odd lemon-coloured things with long hair and even the occasional Siamese (a royal beast in its native land, but when first seen in England referred to as a 'purgatorial nightmare'). Whatever their genes, they all behave in the same self-assured way, which is why London's lawns are a stinking battlezone.

The capital's cats are an experiment in evolution. They retain more independence than do its dogs, for the average dog lover gives two orders an hour to his pet, and the proprietor of a cat just one a day. As William Conway, Archbishop of Armagh, once put it: 'No tame animal has lost less of its native dignity or maintained more of its ancient reserve. The domestic cat might rebel tomorrow.'

Cats are rather less defiant than the good Archbishop imagined. They have depended on man since long before Dick Whittington's day. In Egypt around 2000 BC the animals became gods. When a cat died, its owner shaved his eyebrows as a gesture of respect. Millions of the animals were mummified and, in the nineteenth century, their corpses were shipped to Europe as fertiliser (the auctioneer used an embalmed cat as a gavel).

On the farm, cats are not decorative but useful. The fine for killing one in the days of the Welsh monarch Hywel Dda was enough corn to cover the corpse, suspended by its tail with its nose on the floor. Cats still play a part in the rural economy, but have good reasons to prefer urban life. A city may hold two thousand in each square kilometre – which gives London a million cats of its own. In the country, the animals are a thousand times less abundant and even a large farm contains no more than a dozen or so.

Town cats and country cats differ in their genes. In cities they exist in a huge variety of form, but on farms most of them are marmalade cats, with tawny coats. City cats have evolved into such bizarre variety through man's generosity. By feeding their favourites, owners release their pets from a large part of the struggle for existence – and evolution is quick to seize its opportunity.

Some of the forces of natural selection ranged against a mutant feline are obvious. A lemon-coloured beast with long hair is not much good at hiding from an angry farm dog. However, most of the evolutionary pressure has to do with sex rather than with survival. At night, to those most involved, not all cats are grey.

The diversity of the urban population comes from a great release from the struggle for sex in the city. All female cats, wherever they are, go to where the food is. Males, understandably enough, search out the places with most females. On farms, where food (in the form of mice) is always short, females have to live in tiny groups. As a result they can be monopolised by a single belligerent tom. Only the best males can win, and the hapless losers are forced to roam far and wide in search of a mate.

A town, in contrast, has a dozen well-fed females in each street and any urban tomcat, crippled by his genes though he might be, hence has a chance of sex with the cat next door. As a result, every tin of cat food forms part of a truce in the battle of the sexes. Because the struggle for a mate has been so much reduced, town males, with their generous and artificial way of life, are much more tolerant of one another than are their country cousins.

Marmalade cats are common in the country because their coat-colour gene has an unexpected side-effect, for it makes its owners more aggressive. Bad temper is a great asset to a rural male's bleak existence. On the farm, marmalade males always win a fight with the black-and-whites, the Siamese and all other effete urban variants. Only in London can those strange and unnatural beasts make good in the sexual arena and only there can my breakfast-time irritation be directed at such a diversity of animals.

Darwin commented on the 'nocturnal rambling habits' of cats, which mate according to their own rules rather than ours. Such determined promiscuity makes it hard to select breeds of cats as extreme as those found in dogs. Apart from the coat colours, which are easy to identify, cat owners are forced to leave nature to do most of the evolving. Nobody yet has bred a lion-sized feline for the fireside rug (although I would love to see one leap over the fence to attack one of the donkey-sized dogs that daily lay their visiting cards on my step).

What's Your Poison?

The notebooks used by Pierre and Marie Curie are still kept in lead-lined boxes because they are so radioactive, and in the 1940s it was possible for a schoolboy to buy from his local science shop a 'Volatile Plum Pudding . . . it leaves its dish and rises to the ceiling'. Those are just two of the many eccentric facts to be gleaned from Oliver Sacks' much-praised memoir of his youth, *Uncle Tungsten: Memoirs of a Chemical Boyhood*.

My own upbringing was more conventional than his (no eccentric Jewish millionaires) and my father, rather than my uncle, was a chemist (and of soap-bubbles rather than tungsten); but Sacks' remarkable book struck some real chords. Not, alas, of a boyhood spent in the Science Museum or of a deep understanding of the periodic table (which I coped with by chanting 'Here Little Beggar Boys Catch Newts Or Fish' – helium, lithium, beryllium, boron, carbon, nitrogen, oxygen, fluorine and the rest – rather than with the infant Sacks' claimed insight into the structure of atoms), but of a wild disregard of common sense.

I cannot match his anecdotes of visiting the local emporium for cyanide, thallium and hydrofluoric acid (sold in a rubber, rather than a glass, bottle because it was so powerful) but we took plenty of risks. My father – deaf in one ear and with a face scarred by a lab explosion – sometimes brought home strips of magnesium which he would ignite in the kitchen and pour on water to show how the flames shot up from the bubbling hydrogen (the melamine table was never the same afterwards). At school the chemistry lab could be smelled fifty yards away and, although there must have been others, the sole safety instruction I can recall was the ditty: 'Here lies a schoolboy's body, Alas he is no more, For he poured his H_2O into his H_2SO_4' (a reminder that a test-tube of water added to a flask of concentrated sulphuric acid – rather than vice versa – boils and sprays vitriol into the experimenter's face).

The stench came from a leaky Kipps apparatus that made hydrogen sulphide, a gas capable of inducing a headache even greater than A-level chemistry itself. Today's texts point out that 'chronic exposure leads to memory loss, paralysis of facial muscles and blindness', all of which we identified in our teacher (Dippy by nickname and – it must be said – by nature) with no thought of quite why he behaved in such an odd way or of why we felt so ill at the end of each class.

To settle their stomachs after the daily gassing the youth of the sixties could always sample a mouthful of some random poison by sucking too hard at a pipette. The frivolous tried it with mercury (which was too heavy to draw into the tube) and then threw the lethal liquid at their friends to watch it splatter. When that pastime paled, it was time to drop concentrated nitric acid into a neighbour's pocket to sample the joys of a dissolving jacket. For biologists the rite of passage involved choking over a dogfish pickled in toxic formalin. Such fine disregard for safety went on after school and my first job was in a power station, where I was set to stripping asbestos off pipes without benefit of mask. Later, in a biochemistry lab, I splashed gloveless in acrylamide solution with no thought of nerve damage.

Such memories are amusing enough, but plenty of scientists have paid the price for rash behaviour. Today's rules are irksome, but necessary. In schools, Kipps is dead (or locked in a fume cupboard), formalin is out and a worldwide move to get rid of liquid mercury altogether is well under way. Asbestos has been banned and biology students spend their time watching videos rather than – heaven forbid – cutting up corpses.

All this may be inevitable, but the obsession with security can go too far. Nature has its own risks: hydrogen sulphide seeps from sewers and acrylamide is found in fried food (although hamburgers kill by clogging arteries rather than by poisoning nerves). Real worriers should avoid organic vegetables as the fungi found on unsprayed produce can cause cancer, and should lay off tea and coffee for the same reason.

Today's researchers face a more subtle danger, for repetitive strain injury from filling in hazard reports is a real peril. Oliver Sacks' editor is at particular risk, for the most extraordinary fact in his extraordinary book is hidden away in the acknowledgements. The manuscript was, they say, reduced from two million words – twenty solid paperbacks' worth – to its present modest length. Its author may feel safe enough from the typist's curse, for he claims to do his writing in the traditional way, but he should remember that the aniline dyes used in ballpoints may be poisonous when taken in excess. Sacks: stop sucking that pen!

Raising the Bar

Every year the Jerusalem police pick up several Messiahs who bother tourists with their bizarre behaviour. The Dome of the Rock is surrounded by crazed escapees from *The Life of Brian*, some of whom have an unnerving ability to chant in a high falsetto mixed with a sonorous bass. Faced with false prophets, most of their victims flee, paganism unscathed.

The real Messiah sounded, no doubt, much the same. So does St Matthew in Bach's version of his Passion and any number of the singers of sacred works who use the laws of physics to scare or inspire their listeners into a belief in the supernatural. It all turns on acoustics. Every voice depends on a series of waves generated by the vocal cords and on their resonance in the open spaces of the mouth, throat and sinuses. To sing like Pavarotti needs constant adjustment of the shape of the sounding compartment.

Even the finest singer faces the facts of physics. For higher or lower notes resonance moves to a different part of the sound chamber. If it does not (and the job takes practice) the voice will crack. Popeye and the Spice Girls have tiresome voices because all their notes are kept, on purpose or through lack of talent, in a single register.

For the ancient Greeks the term *mousike* referred both to the verbal and to the musical elements of a piece. Even in modern times, the words of an oratorio or an opera can have a sonorous quality. Say 'Agnus Dei' and the 'Ah' and the 'oos' are sounded low in the mouth, the 'Dei' much higher. Verdi's setting of those words does just the same, to add depth to his music. A trip around the vocal apparatus also gives a plaintive quality to such lines as 'He was despised and rejected' from the crucifixion section of Handel's *Messiah*, where the shift to a low register at 'despised' unites the music with the words.

All songsters must sing from the same hymn sheet as the orchestra, in terms of key at least. Its rules are set by the note middle C, the gatekeeper of the musical universe. Like any other note, it is no more than a sine wave whose pitch depends on the rate of vibration. Even so, the great arbiter of musical correctness turns out to be a slippery customer.

In Bach's time, chaos ruled, with its value set at a different pitch for sacred music and for secular. In the days of the great composers (as proved by Handel's own tuning fork, which has survived the accidents of the years), C stood at 256 hertz (256 cycles per second); a figure based on the Greek ideal of the eightfold symmetry of the universe (256 is two raised to the eighth power). For a whole series of musical works, sacred or profane, such a level allows most voices to shift register at a point where, as in the *Messiah*, the message fits the melody.

Nowadays music – like many intellectual pursuits – has problems with standards. Today's teachers, of harmony or anything else, worry about grade inflation, for it has become much easier to obtain an 'A' in examinations. A 'C', once a respectable mark (I have many), has become a badge of shame. Music itself, in contrast, has gone the other way. The note A, at 432 Hz (which gives C at 256) is now set, in most European halls, at 440 Hz and in some opera houses has risen to 450.

Most people have some pitch memory and when asked to sing the Beatles' 'Yesterday' do so close to the original key.

About one in ten thousand has absolute pitch – the ability to produce a particular frequency on demand – and there might even be some genes involved in such a talent. However, a demand for brightness of tone has pushed classical pitch to its present Olympian levels and, with the A of American Concert Pitch set at 461 Hz, many vocalists simply refuse to join in.

All this is bad news for those sopranos on the high Cs who can bring audiences to tears or to religious ecstasy. Worse, it can remove meaning from the music, for it detaches a register shift in the voice from the words themselves. In the new high key, the performer shifts at an arbitrary point in the sentence rather than where the composer planned it. Like *The Life of Brian* compared with the Gospel of St Matthew, form and content lose touch with each other.

Stung by the way in which the hurdles in the musical steeplechase have been raised, many artistes – Pavarotti included – have signed a petition for lower pitch. The State of Indiana once planned to pass a law defining the mathematical symbol pi to be 3.0 rather than the awkward 3.14159 . . . Fortunately for the stability of the universe, it failed; but an Italian senator has now, with rather more logic, introduced a bill to return A to 432 Hz.

Teachers, depressed by endless official interference, can remind themselves that education, too, needs a fit between words and music and that, as in grand opera, higher fences can mean lower standards.

With Fatal Ease

In the course of my career I have written too many books and too few scientific papers. With the advent of the word processor it has become fatally easy to check one's progress. This volume, for example, took 369,639 typing motions, not counting the surplus stuff thrown into the grammatical burial ground hidden within all hard disks. My difficulty is that the second finger on my left hand hurts, thanks to repetitive strain injury. Composition has, as a result, turned into hard labour.

I could clutch at a straw and bring out a work lacking in any wordsmith's most abundant symbol (*Gadsby*, a book of that kind, authorship of which I cannot pinpoint at this instant, starts: 'If youth, throughout its history, had had a champion to stand up for it; to show a doubting world that a child can think; and possibly, do it practically; you would-n't constantly run across folks today who claim that "a child don't know anything"' and so on for fifty thousand words), but I will probably just plug dumbly on until a final paragraph is at last in sight.

No, I surrender, I must (as a touch-typist) use my painful finger each time I press the key for the letter 'e', whatever the price – ouch!

To succumb to fatal ease is bad enough, but think how much worse things are for musicians. They make not thousands but millions of hand movements as they rehearse, and pay for it. Even if your name is Ernest Wright (the author of *Gadsby*) a book without an 'e' is feasible (or practical). However, *The Ring Cycle* without a B Flat would sound most eccentric (odd, idiosyncratic). A musician whose fingers have stiffened up is, as a result, in serious trouble.

Medicine has specialists who deal with the problem. The figures are impressive. One presto by Mendelssohn demands 5595 notes in just over four minutes – with, in places, seventy-two finger shifts a second. Computer simulation shows that the muscles between the wrist and fingertips can produce a billion billion different movements. Half of all musicians report some medical problems and some have to give up altogether because they cannot play on.

Doctors can give some simple advice to reduce the amount of damage – put wheels on your cello case, follow the path taken by the guitarist John Williams and never practise for more than twenty minutes at a time and if you have short arms use a bent flute; but, too often, surgery is unavoidable. Many musicians are terrified of the idea. Schumann, in contrast, turned to a surgeon to make his fingers easier to bend – and it destroyed his career and perhaps drove him to his final madness. Sometimes, though, the knife may be unavoidable.

Bagpipe players have a special problem. They use the middle segment, not the tip, of some of their fingers to stop the holes and, because that part of the digit is not very flexible, they face real difficulties. Much worse, many Scottish pipers suffer from the Curse of the MacCrimmons.

That grisly condition is named after the great family of pipers to the Clan McLeod of Skye, composers of much pibroch music, the staple diet of bagpipers. Many of their descendants carry on the musical tradition – and the MacCrimmon gene for a hereditary condition called Dupuytren's Contracture. It starts after the age of fifty, with nodules in the palm. These form bands of fibre that run into the fingers, causing them to close up. For most people the little finger, crucial to a piper, is the first affected. The surgery involved is somewhat wince-inducing (triangular skin flaps are much in vogue), and the best to be expected is five or ten years of relief before the symptoms start again.

Some pibroch players face even worse difficulties. One piper cut off his thumb with a circular saw while at work on home improvements (a dangerous hobby for any musician). The stump was too small to let him play the chanter of his instrument. The solution was bold but simple – cut off his big toe and graft it in place of the missing thumb. Three months later he was playing his bagpipes again.

The operation to make a piper who plays with his toes has been done only once, but speaking as a man who, as a student, lived for several years in a shared apartment within hearing distance of bands practising on Edinburgh Castle I forbear to ask 'How can they tell?'

During my sojourn that Midlothian flat was a sanctuary for clinicians from psychiatrists to otorhinolaryngologists. An occupant from long ago, now a famous hand-doctor, Ian Winspur (with his collaborator C. B. Wynn Parry), spins that bagpiping yarn in a book about musicians' hands put out by Martin Dunitz and Company, of London. I can, thanks to a lack of a crucial and usually abundant typological symbol in this last paragraph of my discussion, finish this short account of surgical nostrums for musical Scots with no additional strain on my own painful digit.

We Shall Fight Them with the Telemöbiloskop

Who invented radar? We Brits are quite confident; it was those noble Anglo-Saxons Messrs Boot, Randall and Watson-Watt (with perhaps a little help from the Americans, who in fact spent more on developing the device than on the atom bomb).

Not everyone has such an Anglocentric view. Sixty years on from the Normandy landings, the French came up with another idea. *Le Monde*, that pillar of Gallic rectitude, pointed at a different source. In 1934 Emile Girardeau patented a system that was, the paper claims, the first effective radar ever used. Most important, his device used short waves – sixteen centimetres in wavelength – while as late as 1940 the best that the British could do was to make waves ten times longer. As the acuity of a radar device depends on the distance from the peak of one wave to that of the next in relation to the size of its target, Girardeau's invention could have been crucial in the first days of the war. Instead it was ignored by his own military men (although, with a certain irony, it had already been installed as a safety feature on the liner *Normandie*).

Today radar is everywhere. Its waves come in many lengths and are aimed at a variety of objects. Some are in the metre range used by the first British wartime kits, but rather than searching out aircraft they are used to hunt for larger items in the sky. Already such machines have found more than two hundred asteroids around our planet. Most are revealed by the bouncing beams to be no more than piles of rubble held together by gravity. Some are long and thin, and tend to split into two to form a paired structure when they get too close to the Earth's gravitational field.

For those worried about the possibility of renewed aerial attacks the trackers warn of a blitz by a kilometre-wide object on the afternoon of 16 March 2880. Asteroid 1950DA was first spotted five years after Armistice Day and has been followed ever since. The boffins are not entirely certain whether it will strike, because the object's path is influenced by the energy it soaks up from the Sun. That in turn depends on its exact shape, how much the great rock spins and on the colour and texture of its surface. Radar technology is, alas, not yet quite good enough to measure those. To reassure anyone already on the way to the Anderson shelter, the odds of a hit are at present estimated to be three hundred to one against.

At the other end of the electromagnetic spectrum, too, things have moved on since D-Day. Millimetre-wavelength radars use a compact antenna to generate a narrow and well-focused beam that can search out small and distant objects. They are much utilised to study the movements of raindrops in clouds, which is useful for weather forecasters. The machinery is light enough to fit on a small satellite, and such devices can identify ocean swells and even ripples. Given the worries about wind and waves at the time of the Normandy invasion, they would have been valuable indeed six decades ago.

Many civilian advances in technology have been pushed by military needs. Today's millimetre radars say exactly where a hostile tank might be and how fast it travels and can even record the size, direction and muzzle velocity of the bullets that stream from its guns.

The Allied invasion forces did face a detection system as they crossed the Channel (and much effort was expended on the day in dropping an advancing wave of aluminium foil to fool it into thinking that a phantom fleet was on the move). It was based in the main on the German Würzburg radar apparatus, which by 1944 was hopelessly out of date. With a wavelength of fifty-three centimetres a Würzburg needed a huge and vulnerable antenna and was not at all accurate. It faced much smaller and more precise machinery on the Allied side, some of which, even in those early days, used a wavelength of just a centimetre.

The Germans' technical failure of 1944 is a surprise, for they had got there first, many years earlier. In 1904 the German Patent Office granted rights to the twenty-two-year-old Christian Hulsmeyer for an invention he called a 'Telemöbiloskop'. It was the first radar, an apparatus that sent out a fifty-centimetre oscillation and picked up reflections from passing ships. It would, he thought, be useful in preventing collisions, but the notion did not catch on. The idea was developed to some extent by the Nazis, but Hitler did not have much interest as its prime use was in defensive warfare and it would never be needed . . . would it?

Evolution on the Costas

Biologists nowadays have less and less to do with biology. I went into the trade, I admit, because of a childhood interest in bird-watching. Two grim years of chemistry, physics and maths at A-level, I promised myself, then the animals, the plants and what I thought was real science at last.

The real thing, forty years ago, was a nasty shock. Introductory biology at university was a collection of lists: of insects, of worms and, once in a while, of things with fur on them. Sometimes the lists were attached to ghastly objects in bottles, but most of the time they were just lists. Science was not much more than stamp collecting – and the more creatures you could name, the higher your mark.

Now things have changed and, generally speaking, for the better. Today's students do experiments that were at the edge of technology a mere five years ago. What's more, they have the gall to be bored by it. I was amazed when I first saw DNA. Modern undergraduates say, 'Not that again, we did it at school'! But, safe from rote learning, they have lost part of their education. To do anything decent in science, it seems, needs machines and money. To make a scientific catalogue today, whether of yet another atomic particle or a few hundred million DNA bases, is expensive and difficult.

Every spring I get a reminder that science need not be like this. For many years my colleagues and I have run a field course in southern Spain. It seems odd to do genetics in the open air – but Darwin did just that. If nothing else, the course is the southernmost educational experience in

Europe, set near the town of Tarifa, just ten miles across the strait from Morocco.

To reach it from the airport one drives through the Costa del Sol, Croydon with mountains. Sprouting among the rubble and the English pubs are the disregarded remnants of what was once a beautiful ecosystem. And, west of Gibraltar, most of it is still there.

Each year about thirty students come for ten days. In that time we make at least a dozen decent little discoveries. None is spectacular, but they are interesting enough and some reasonable science is done on the way.

Are you aware that the slightest bit of asymmetry in a flower destroys its attractiveness to pollinators? Perhaps they sense that if its genes are not good enough to make a decent flower the plant will not provide a fair reward and will fail in its evolutionary task of passing on its biological heritage. I didn't know that – and neither did anyone else – until the students found it out a few years ago. Such work had been done with much bother and expense on birds, but it is nice to see that the same rules apply to flowers, too.

And those paper-wasps, the ones with elegant little nests stuck to leaves . . . Some nests have one wasp, others many. To grind up the insects and look at their proteins shows who is related to whom. All wasps in a nest are, the genes show, close relatives; in other words, cooperation demands kinship. Within a nest one of the wasps is boss while the others do all the work. Put a stranger in and the locals attack. The more different its genes, the sooner it is killed, but soak it in an extract of one of the family members and the invader is allowed to join the group. Again, the result will not change the world, but it is not bad for three days' research.

Life is not always so rosy. Often it rains for days on end. As despair sets in I remind myself of Darwin, whose life-long memory of his three-year trip on the *Beagle* was 'one continuous puke'. It does not help.

Other crises can happen. Once a student was attacked by a Black Kite – an impressive bird at close quarters – when it mistook her brown hair and red ribbon for a juicy rabbit. On another trip a hapless lad disappeared just as a drowned body was found on the beach. We found him asleep behind a bush.

But, for most of the time, such horrors seem worthwhile. Yes, there can be dreadful hours of statistics. And there are endless arguments as students make the most important discovery in science – that someone will always try to steal your results. It is cheap too, for it costs less than an equivalent field trip in England, and the price is trivial in comparison with that of the chemicals poured down the sink by molecular biologists.

Most important, it reminds me of a truth too easy to forget – that science can, sometimes, still be fun.

What's in a Name?

One of the more obscure statements of Marshall McLuhan (media, messages and so on) is that 'the name of a man is a numbing blow from which he never recovers'. From my own perspective I see what he means. The sole Jones in central London, in the way of thoroughfares at least, is attached to a tiny alley off Berkeley Square. Compare that with Smith Street, Square, Terrace and Field in the centre and twenty more out in the suburbs; but at least my alley is unambiguous. And when it comes to Duke Street (Mayfair) and Duke Street (St James's), less than a mile apart, no wonder the taxi drivers get confused.

The map of the genes is a lot more complex than is that of London; but vagueness, ambiguity and chaos rule. The American journal *Science* makes an annual award to the Breakthrough of the Year. Recently it was, as many expected, the protein p53. Much fancied, but pipped at the post, were ApoE4, p91 and SOS. Who are they – and what do they do? Their names certainly do not give much of a hint.

In spite of its tedious title, p53 is remarkable stuff. The protein is a field marshal in the defence against cancer. It slows down cells whose division has gone out of control and even persuades those errant individuals to commit suicide. When p53 goes wrong, though, cancer begins to win. Mutant mice without the gene always die within six months. Several human cancers are due to errors in p53, and smokers may wish to note that tobacco is an excellent way to damage the magic molecule. p91 (no relation) is part of a system that sends messages around cells, while ApoE4 is involved in Alzheimer's disease. SOS is in yet another cancer pathway and has no tie with shipwrecks.

Hearty congratulations, then, to p53. But the journal's choice shows a certain lack of imagination. Any scientist who reads papers outside his own field is at once blinded by acronyms. Most people can manage DNA (although I once met a policeman who mixed it up with DOA – Dead on Arrival). But (from a recent issue of *Science* itself) do we really need ω-Aga-IVA, let alone [³H]GDP-Cdc42? Such names are writ in water, which itself came a poor fifth in the *Science* sweepstake, in the form of Greenland ice with holes drilled through it to look at climate change. No one can remember what the acronyms mean for more than a few minutes.

To avoid such obscurity, fruit-fly geneticists have gone to the opposite extreme. The thousand-page list of variants in *Drosophila melanogaster* (the black-bellied dew lover, in literal translation) has lots of venerable and rather insipid examples such as *White Eye* and *Vestigial Wing*. Nowadays, though, the Gothic tastes of those who christen the hundreds found each year have started to run out of control. The *Swiss Cheese* mutant has holes in its brain. *Lot* flies enjoy the taste of salt. *Shy* animals never emerge from the egg, while those with the *Stoned* gene behave in a rather odd way. Other intellectually challenged flies include *Turnip* and *Cabbage*.

Sometimes the names are more eccentric than informative. What is to be made of *Killer of Prune*, *Faint Sausage* and *Ether a-go-go*? *Sonic Hedgehog* is named after a cartoon character; and because it has become important in developmental biology, hundreds of people write learned papers in which that unlikely individual stars.

Now that complete gene maps of dozens of species are emerging, what was a mess has become a nightmare. The *Armadillo* gene gives a *Drosophila* larva an armour-plated appearance; but DNA shows that this is in fact exactly the same as the mouse gene called beta-catenin. Which is the real name and which the alias?

Now the science of gene ontology has (with the help of a P41 grant from the Human Genome Organisation) started a desperate attempt to sort out the chaos. Even within humans themselves, with their twenty thousand or so genes, the task is not easy. One human gene (SELL, to its closest friends) has been given fifteen different names by different research groups, and one name, MT1, has been attached to eleven different genes. To sort it all out may take years.

Perhaps the tobacco companies could sponsor the ontologists in return for being allowed to give certain genes a title in keeping with their job – *Silkcutene, Marlborase*, even *Woodbine*. They might even spend some money to persuade people to check that gene p53 is in good order before they light the next cigarette.

Chaos in the Heavens

Life in this twenty-first century has been less than predictable. Faced with political turmoil, many have turned to the heavens for advice. Nostradamus saw the future in the stars and Isaac Newton himself imagined the cosmos to be a giant clock wound up by God. Once dismantled, it would reveal the Creator's plan. His own great contribution was to simplify the universe; to show that previous notions that stars and planets loop crazily through the sky were too complicated. The Sun was in charge, we were inhabitants of a mere planet and the solar system could be reduced to a mathematical formula.

In 1999 the Earth passed through the Leonid meteor shower, a belt of stellar gravel that swings past the Earth every thirty-three years or so. Its previous visit, in 1966, put on a great show (albeit not as great as 1833, when Americans were thrown into a panic by an earlier rain of fire from the heavens). In the weeks before the latest influx astronomers made their forecasts. The prediction for Europe was no better than: 'this shower could reach storm levels, or could no-show'. But why the uncertainty? Thanks to Newton most of us are confident that there is rather more than a 'good chance' that the Sun will rise tomorrow morning. What makes the meteors so much less dependable?

The vagueness of the astronomers sounds like a typical Nostradamus prophecy: 'The year 1999, seventh month, From the sky will come a great King of Terror, to bring back to life the great King of the Mongols.' An augury of the

Twin Towers by the great clairvoyant of the Middle Ages, perhaps – but July 1999 is not September 2001 and predictions are not much use unless they arrive on time.

Nostradamus and his modern fellows could not survive without a certain diplomatic vagueness, but surely astronomy must be more precise. Why are its experts unable to say just where, when and how abundant the great Kings of Terror, the Leonids, will be?

Newton was keen on astrology but his successors are, in general, not. However, his formulae make sense of the heavens only up to a point. The laws of gravitation are simple enough for a sun and a planet, but once more than two bodies enter the equation life gets far more complicated. Nineteenth-century mathematicians got stuck with just three objects, Earth, Jupiter and the Sun. Try as they might, they could not fit their behaviour into the Newtonian rules, for mathematics predicts that the lightest partner (our own home) should bounce around the solar system, which, fortunately for those who live there, it does not.

The many-body problem, as it is called, hints that the universe is less rational than it seems. As more and more objects, from planets to asteroids, enter the equation and exert gravitational force upon one another the sums become fierce and, in effect, insoluble. Quite soon the rational world of Sir Isaac breaks down.

Astronomers have been forced to move on from his notion of the universe as an ordered and predictable system. Chaos has entered the heavens; and, for many of its denizens, what they will do tomorrow depends in a quite unpredictable way on what they did today. Chaotic systems, the universe included, behave in a calm fashion for a time, but sometimes break into pandemonium.

Meteors are influenced by many gravitational fields. They include that of the Sun, those of the nearby planets and the much smaller contribution of their fellow cosmic pebbles. As a result, the outer asteroid belt loops through the heavens under the influence of Jupiter and Saturn and does so in a manner that simply cannot be predicted. The mathematics hint that such cosmic elements may even spin round one star for millions of years, to be hurled with no warning for immense distances to settle around another. Even over a year or so it becomes difficult to tell what a meteor shower will do next (which explains the uncertainties that surrounded the 1999 storm forecast).

Chaos also rules for more substantial members of the world above our heads. Outside our own solar system certain planets have rocketed from the influence of one sun to another as their orbit became unstable. Closer to home, the orbit of Mercury shows alarming wobbles that could break off into chaos some day (probably, but not certainly, not tomorrow). If Jupiter were any larger than it is, the Earth itself would be in severe danger of being thrown out of the planetary club at some quite arbitrary moment. The distant giants Jupiter, Saturn and Uranus form a three-body system and are liable to such behaviour (although mathematicians estimate that Uranus should stay around for another billion billion years before it zooms into the void, which should see most of us out).

In the end, Britain's attempt to admire the Leonid storm fell before fickleness; not of the higher mathematics, but of the weather. We were promised a clear night, but in the south it rained, and there was nothing to be seen. Better luck next time in 2032, possibly.

A Sensitive Prince

Charles Darwin, that prince among scientists, spent long hours playing a bassoon to his plants. Charles Windsor – a prince among, well, princes – would no doubt approve: what could be more in harmony with nature than to eat organic oatmeal biscuits to the sound of a wind ensemble? And if one can hug trees, one might perhaps with equal profit serenade them in the same way.

Darwin ('evolution' – explains our prospective future ruler – 'is a man-made theory to explain the origin and continuance of life on this planet without reference to a creator') was not in search of his inner self or that of his leafy friends. Instead he was carrying out an experiment (a pastime not much appreciated by the heir to the throne). It led, oddly enough, straight from the greenhouse at Down House, Darwin's home in Kent, to the world of high-tech agriculture so often deplored by Prince Charles.

Darwin's 1880 work *The Power of Movement in Plants* looked into how some climbing plants twist, how others manage to close their leaves at night and how a few (such as the sensitive plant *Mimosa pudica*) respond by folding their leaves when touched. Light, he found, had an effect on some of the patterns of movement, but loud as the bassoon might be, his subjects remained deaf to its sound.

He never found out how plants respond to gravity, light or touch, but soon it became clear that chemical messages – hormones – were involved. Dozens have been found. In both plants and animals such internal messengers do their job when they bind to DNA and switch particular genes on or off.

Not long ago botanists returned to the question of the sensitive plant. The experiment was simple: take a series of plant hormones, water the subjects with them and see which bits of DNA respond. For a surprisingly wide variety of such chemicals, a previously unknown gene increased its activity by a hundred times.

All very satisfactory, but every experiment needs a control, which in this case involved sprinkling the plants with water alone. To the scientists' surprise, this too activated the new gene. Hormones had nothing to do with it. In fact simple contact does the job. A drop of rain – or a gust of wind – and certain genes (the 'touch' genes) spring into action. Among their other jobs, they slow a plant's growth when activated, which is why a wind-battered tree on a crag above Balmoral is smaller and more gnarled than one from the forests of Highgrove. The researchers, in homage to their nineteenth-century hero, also tried loud music on their plants (although they do not reveal whether it was heavy metal or Richard Rodney Bennett's Concerto for Bassoon and Strings) and found, with Darwin, that they remained unmoved.

The touch genes opened the door to a great family of relatives. Their products act on the plant cell wall and cause it to firm up or loosen its fabric in response to stress. Growth patterns and the extent to which cells stick together alter as a result. Some members of the clan even respond to the passing seasons (which is why leaves fall off in autumn).

The proteins involved send signals within the cell and are similar in structure to the molecules called calmodulins that do the same job for us. Calmodulins are internal messengers, proteins that sense the level of calcium in the cell and pass the information on to the variety of enzymes and membrane channels that respond to it. Among other tasks, they control our heartbeat, switch on hormones that determine growth and, for that matter, alter the blood chemicals that change our mood from happy to depressed.

Touch genes may soon be engineered to give fruits that fall from the tree in a slight breeze as soon as they are ripe or plants that grow tall even in windy places. The green move-ment would no doubt disapprove for, they say, all this – to quote our prince – just reduces the natural world to nothing more than a mechanical process.

But is there not something magical about such layers of scientific rationalism – about how calmodulins connect bas-soons with bonsai and the patter of rain with the beat of the heart? Why turn to mere romance in the vain search for a guiding hand?

Plenty of romantics agree. Shelley writes of a garden in which a mimosa droops in response to a rejected lover's despair: 'Whether the sensitive Plant, or that / Which within its boughs like a Spirit sat, / Ere its outward form had known decay, / Now felt this change, I cannot say.'

Biology will never answer Shelley's question or others like it, but the poet himself saw that science told us things about the world that poetry cannot. He filled his Oxford room with crucibles, chemicals and electrical gadgetry (which hints at why his sister wrote a book so useful to foes of Frankenstein foods) and saw no contradiction between the world of the spirit and the universe of science. He would, no doubt, have been delighted to learn that cooling passions are indeed linked to falling leaves.

And might the Prince be dissuaded from his mystical meanderings by the discovery that if you hug a tree you stunt its growth?

Diamonds Are For Engels

Much scorn has been heaped upon the United States' electoral system, which in 2000 allowed a few hundred votes to change the fate of a nation. The Electoral College, its representatives chosen by each state, was the target of particular criticism by Bush's opponents. How could Florida's margin of victory of a few tenths of a per cent be translated into twenty-five College votes, all for George W. Bush, a result which altered the complexion of the whole country?

It seems unfair, but the idea was supported by Karl Marx (a democrat if not a Democrat). In *Das Kapital* he noted 'that merely quantitative differences beyond a certain point pass into qualitative changes'.

Marx was concerned with politics, with how much property a man must have before he is transformed from sooty proletarian to diamond-encrusted capitalist. His colleague Friedrich Engels, in his discussion of Marx's ideas, saw that the law of transformation of quantity into quality went further. In his search for a science-based politics he turned to the chemistry of carbon. In *The Dialectics of Nature* Engels points out that as the number of carbon atoms increases, chemicals in the paraffin series shift from a gas (as in the one-carbon methane) to a liquid in those with five to twenty carbons, and to waxy solids above that. He noted that water, too, shifts from ice to fluid to steam at two crucial temperatures. A tiny change in the right place is enough to transform the state of a chemical. Florida, in fine Marxist tradition, did the same to the chemistry of a state.

And that is the point of the Electoral College. It leads to

what chemists call a phase transition: an unequivocal shift from one configuration to other – gas to liquid, or Democrat to Republican – when a critical point is reached. To get through the barrier, in either direction, needs work (which means that a phase change, once made, is hard to reverse). The Founding Fathers, who invented the system, saw to it that American politics is at least decisive. The positive feedback they built in causes a tiny margin in a state's popular vote to be transformed into a large one in the nation's Electoral College and to an absolute difference when the President is elected. The College is, in chemical terms, the transition state, a point that is of its nature unstable and bound to tip one way or the other.

The same is true of Engels' element. Carbon can arrange itself in several ways. Charcoal, a shapeless mess, is one, and graphite, soft because its atoms are arranged like sheets of chicken wire that slide over one another, is another. One form common in Florida is diamond, whose carbons are arranged in a regular lattice, and yet another is buckminsterfullerene, found in soot, in which sixty units make up the form of a football.

The chemistry of all four substances is just the same, but the phases are so distinct that they seem almost unrelated. Even so, just as in politics, under the right conditions and with enough effort the ruling form can be displaced. Diamonds are made fifty miles underground, at enormous pressures and high temperature. In 1880 the Scot James Ballantyne Hannay claimed to have manufactured a few by heating paraffin in a sealed tube (an experiment that caused uproar in the markets, but could not be repeated).

Since the 1950s diamonds have been synthesised from graphite, heated to 1500 degrees at sixty thousand atmospheres of pressure. More are manufactured in this way each year than the total ever mined. They can also be made from buckminsterfullerene at room temperature and a mere twenty thousand atmospheres. Even the carbon held in methane can change its state to diamond, for when the gas is passed over a hot filament in a vacuum it shifts phase straight to the ordered crystal found in engagement rings. The chemical pendulum can also swing the other way. In 1772 Lavoisier burned a diamond with sunbeams focused through a giant lens (another experiment not often repeated).

Why, then, are not diamonds broken down into graphite – or buckminsterfullerene – as they explode from the depths of the Earth? In fact most of them are, but in some places they shoot out with such violence that they cool too fast to allow the phase transition to take place.

To shift any substance from one state to another takes energy: heat and pressure to change carbon; dollars when American politics are involved. The presidential election is a system pumped full of cash and often trembles at the transition point. Those worried about its apparent unfairness should remember that just three countries in the world elect presidents without some form of intermediary body: Finland, which has no politics; France, where the normal laws of nature do not apply; and Russia, where Boris Yeltsin once won with a third of the vote.

Evolution as Thatcherism

In the first *Planet of the Apes* film (which I saw in Bucharest in 1968: nobody laughed – but Ceauşescu had just come to power) the apes become human in their behaviour but, in appearance, remain chimps. For Hollywood the arena of evolution had shifted from body to mind and from Africa to the USA.

That notion has some truth, for *Homo sapiens'* recent past is quite different from that of its primate kin. Should a chimpanzee sit next to me on the Tube I would find another seat; but if an early human from a hundred thousand years ago did the same I might not notice. Men have altered rather little in their bodies since they began, although mentally they have shifted beyond recognition. Their relatives, in contrast, have evolved a lot. A single chimp social group in West Africa contains as much genetic diversity as the whole human population – and the difference among groups a hundred miles apart is greater than that between Aboriginals and Icelanders.

Man may be the primate who did not evolve, but he is driving his fellows to extinction. When future palaeontologists dig up today's fossils there will no longer be a dense bush of primate lineages, but just one human branch. As a result the Earth, given the crisis facing our wild relatives, will soon be – alas – a mere Planet of the Ape.

Chimpanzees do not look to the future. We do, and it will probably be rather like the past. Our gene pool has been pretty untroubled since Cro-Magnon times, and is becoming more so. Evolution depends on differences in the ability to copy genes. If everyone had two, or three, or thirty, children, it could not happen; natural selection can work only when some people have, for genetic reasons, more progeny than others.

In the old days its mighty machinery had plenty of raw material. In Glasgow's magnificent Victorian cemetery, the Necropolis, the height of each monument (and some tower over the visitor) shows how wealth and death go together, for the bigger and costlier the stone, the longer the grave's occupant lived – and, the inscription shows, the more children he had to share his grave. Many of the hungry Scots poor had no tombstones (or offspring) at all, and their genes died with them.

Glasgow in its injustices is still a factory for Darwinism. There poverty and wealth live close together, with differences in life expectancy of ten years within the city limits. Even so, the class differences are smaller than they were. No longer do a few people leave a dozen descendants and a great number none. Even Glaswegians are becoming more and more alike in the chances of survival and of reproduction. As a result, the rate of Caledonian evolution has gone down.

Natural selection's coffin has not been entirely nailed down. The United Nations has, without realising it, begun to measure the power of Darwin's machine. Because relative genetic success is related to social difference, its economists can predict where human evolution will happen.

The Gini Index is a statement of a nation's divergence in income between top and bottom. The larger the figure, the more the inequality; and the greater the variation in life expectancy and in the numbers of surviving children per citizen. At Gini Zero, all families have the same income and more or less the same number of offspring – which means no chance for natural selection. At 100, the Gini maximum, one family has all the wealth (and, presumably, all the children) and the Darwin machine proceeds at full speed.

For Gini the global picture is mixed. Most of Europe is an evolutionary dead end, with too much equality to be of much interest. Denmark scores twenty-five and Slovakia even lower. Britain and the United States have benefited from several decades of political progress and click in at thirty-six and forty, with no sign of a decline in recent years. The global centres of inequity are in South America, where some countries have a Gini score of over sixty, and in Africa. The economists have pointed to where human evolution will happen; and most will, as ever, take place in the tropics.

There is one exception. The collapse of communism has done a Gini to Russia (where inequality has rocketed almost to African levels). There, things have got worse since *The Planet of the Apes*. The next film in the series might feature not the ruins of the Statue of Liberty, but the wreck of the Kremlin. Bucharest, alas, has no monuments worth preserving.

A Face in the Crowd

Painting, said John Constable, is a science, 'a branch of natural philosophy, of which pictures are but experiments'. He had a point. A whole science asks why we are so good at telling faces apart and so keen on portraits. Two scrunched-up newspapers look much the same although their shapes may be quite different, while two faces are distinct although their shapes are almost the same. The camera in the brain can recognise people with remarkable accuracy.

At least, it does most of the time. I once played a minor part in a television show about grammar-school boys. It had pictures of 'then' versus 'now' (not, in my case, a flattering comparison). The producers, as they do, sent out a card to announce the date of the show; a reduced version of one of those vast group photographs, with pupils in their hundreds, teachers in the favourable ratio of the 1950s, and gowns, blazers, ties and the glum faces of those days in abundance.

Ah, what memories! I could easily spot my class at Wirral Grammar School, the flower of Liverpool's Left Bank. There they all were, Anyon, Backhouse, Baker, Beck, Bunn lined up in order, to Williams via a twelve-year-old Jones. What, I wonder, became of them?

But then came a surprise. The older boys in the picture wore striped jackets, which we did not. A second look showed that it was not Wirral Grammar at all but another (and, from the stripes, rather snottier) institution. I had been fooled into seeing people who were not there.

And that is where the science comes in. To recognise a face is not simple. As artists have long realised, it depends on scale, on position and on context. A simple bar code, the position of six stripes of dark and light – hair, forehead, eyebrows, nose, lips and chin – stores most of the data in a visage. The school photograph was just too small to do the job. A school photo is also harder to identify than is a portrait seen slightly from the side, because the latter contains three-dimensional information (the shape of the nose, a furtive ear) which is absent from a full-face picture. In the modern age of paranoia, with scares about terrorism and asylum seekers, vast sums are poured into improving the technology, with plans for facial scans to be included on passports, identity cards and security badges. The technology scarcely works, but the idea buys votes and costs money and industry is glad to oblige.

Some people – perhaps as many as one in fifty – are no good at faces. Instead they use clues from voice, clothes or place to tell individuals apart. The condition may be caused by stroke, or other brain damage, but one form runs in families, with perhaps just a single gene involved.

One such unfortunate was engaged in a lawsuit. At the Old Bailey he discussed his case with a barrister – but the wrong one; not his own, but his opponent's. After all, the context was right: a lawyer, with a gown, in a courtroom. Only the face did not fit. Needless to say, he lost. I, too, assumed that a school photograph must be of my own school and so strong was that idea that it led to false perception.

The law once put great faith in 'photofits', pictures composed of cut-up mouths, noses and eyes. They are unreliable because, like the careless and faceless plaintiff, they are set firmly in criminal context and – like a school photograph – are seen square-on. That leads at once to injustice. If an innocent person resembles a photofit built from witness description, he may be doomed just because a picture of someone who looks a little like him has been put together by the police. That may impute guilt to the innocent – even if the image bears no resemblance to the real villain.

Faces bring art and science together. So strong is the urge to find them – and a baby will respond to a facial cue within minutes of birth – that we see a human face on the Moon. Foolish people have persuaded themselves of the existence of the same thing in the *Voyager* photograph of a mountain range on Mars, as proof of an extinct race trying to send us a mute message. Once NASA published better images from the planet, the mountains just looked like mountains once more. If the credulous were mugged by the Man in the Moon, would they pick him out in an identity parade?

A Golden Wedding

On the beaches of the Pacific island of Yap lie a number of huge limestone wheels each weighing several tons; grim, immovable and apparently useless. Many years ago they were shifted to their new home, with infinite labour, from the distant island of Palau – but why did ancient canoeists make such a risky journey to so pointless an end?

The stones were the basis of the islanders' monetary system; each was a statement of opulence. They proved the financial probity of their owners. Yapese logic might seem odd but is not much different from our own. We take not rock but metal as a metaphor for wealth – gold: heavier than marble (a cubic foot weighs half a ton), equally impractical and even more painful to obtain. Gold bars and massive wheels are difficult to counterfeit and hard to move. They last, more or less, for ever. Another attraction lies in their uselessness. Gravel or steel would be no good as a fiscal reserve because they are needed to build houses, but stone discs and heavy ingots have few other calls upon their time.

Gold is a noble metal. It reacts with almost no other substance – which is why (unlike steel) it always glisters. Its aversion to mating makes it an aristocrat among elements (although, like its human equivalents, it is liable to adultery, a sin tested in Britain by the annual Trial of the Pyx, which each coin must go through to check whether it has mated with an inferior metal). The element does sometimes lower itself by combining, reluctantly, with a more humble substance. *Aqua regia*, Royal Water, is a formidable mix of nitric and hydrochloric acid which can dissolve gold by forcing the metal into a compound with chlorine. Sodium cyanide, too, tempts the precious substance into a relationship (and is much used in goldfields), but beyond that gold likes to keep to itself.

Modern chemistry has pushed it into another marriage; not with some plebeian element, but into an unrelated branch of nature's upper crust. The gas xenon is even more careful about whom it combines with than is King Midas' metal. Xenon is almost inert and has almost no practical use (although the bright blue headlights of modern cars do utilise a bit of it). Because there is no space for intruders to shoulder their way into the outer shell of electrons that surround its atomic nucleus, xenon is almost inactive when faced with other chemicals. It was not persuaded to undergo its first chemical reaction until the 1960s.

It has now made an unexpected liaison, in a remarkable compound of gold and xenon. The chemists who synthesised the new material had made a mix of gold and fluorine in a xenon atmosphere, but to their amazement the noble metal and equally noble gas spurned the planned but plebeian match and fell into each other's arms. Their offspring was a dark red crystal, the most aristocratic substance ever made. Four xenon atoms sit around a gold atom at the core. The structure looks rather like a minute version of the British State Regalia, itself a statement of national wealth.

On Yap, stone represented such a solid financial base that one family gained status by their supposed ownership of a disc that had fallen overboard and sat, unseen, on the bottom of the sea. John Ruskin tells the tale of a shipwreck in which a merchant strapped his bag of coins around his waist and jumped into the sea. He asks: 'Now, as he was sinking, had he the gold – or did the gold have him?'

That question is familiar to economists. At one time the answer seemed obvious: man was in charge, and he who had gold – rare, constant and locked away – had power. Now the world's financial chemistry has changed.

Once trade turned on no more than goods; on cattle, or spices, or (as in post-war Germany) cigarettes as the medium of exchange. None lasted, because they are soon used up. The value of gold derives from its stability. The system worked fine as long as everyone believed in it and there was a shortage of the raw material. After the war more and more poured out of the ground, and the US Federal Reserve sat on huge quantities. Then, in the 1980s, it lost faith, the speculators sold out and the price of the metal plunged. It has been replaced by paper, or electronic signals, and global finances are based on transient symbols rather than nature's elements.

Or are they? Many governments still hanker after something more, well, *real* as a measure of wealth. De Gaulle made a disastrous attempt to return to the gold standard, and the planned sale of four hundred tons of the British equivalent led to uproar as the metal staged a rally after the deal was done. What we really need is something rare and useless as a base for the world's finances – and what better than the new super-noble red crystals? As they break down as soon as they are taken out of the deep freeze they could even help to control inflation.

The Triumph of the Weeds

The Galápagos authorities have declared Santiago Island pig-free. The island has returned to the swineless state it enjoyed when Darwin visited in 1835 ('Mr Bynoe, myself and our servants were left there for a week . . . we lived entirely on tortoise meat . . . the young tortoises make excellent soup'). In those ponderous creatures Darwin saw not just his lunch, but the first hint of evolution, for animals from different islands were subtly distinct. In a rare conjunction of taxonomy with gastronomy, he noted that the Santiago specimens were, compared with others, 'rounder, blacker, and had a better taste when cooked'. (Darwin did better than the Victorian guard cartooned in *Punch* who tells a passenger: 'Cats is "dogs", and rabbits is "dogs", and so's parrots; but this 'ere tortis is a insect, so there ain't no charge for it!')

A century later the hungry naturalist could have chosen a more conventional diet, for Santiago was overrun by pigs. They are among the worst of the island's many alien plagues. Like Darwin, pigs appreciate the taste of young tortoises, and in several places in the Galápagos have driven the animals to extinction. The Charles Darwin Foundation has a tortoise nursery, but without a natural home the future of the Santiago version was grim.

Drastic measures were needed and military technology was called in. Teams of hunters were equipped with Global Positioning System receivers. They cut a grid of tracks through the dense scrub to ensure that the pigs had nowhere to hide. Each hunter moved forward, to form a disciplined

line, shooting as he went. Baits monitored the porcine presence and at last, it seems, the pigs have gone. Already
ground-nesting birds such as rails and petrels have
increased, as have the local snakes (a tasty snack for a hog).
Soon the authorities will be able to leave the tortoises to
breed without help and their electronic sights will soon be
fixed on the goats of Isabela.

Although 95 per cent of the species found by Darwin
still survive, pigs are but one of the threats to evolution's
workshop. The latest Galápagos invader is less conspicuous, but has piggishly broad tastes. The cotton-cushiony
scale insect was first noticed on the archipelago twenty years
ago. It has reached fifteen islands and attacks dozens of
native plants, some of which face extinction. Among the
most threatened are the local mangroves, unique to the
Galápagos.

Why are pigs and scale insects so dangerous? The problem is their Epicurean tastes, for they will try anything once
and – like Darwin himself – turn to the nearest source of
food if their usual diet is unavailable. As a result they can
snack on the last specimens of an endangered species without any danger of eating themselves out of house and
home. They just move on to the next victim. Such predators
are great threats to diversity all over the world. Cats are a
plague; domestic pigeons have pushed their Galápagos relatives out; and alien wasps have done terrible things to the
archipelago's insects.

We are in the midst of a global triumph of the weeds,
and of an era in which narrow specialists have given way to
the vulgar generalists who feed on them. The loutish
strangers, from pigs downwards, have been attacked in
many ways. The buzzword is biological control, but it does

not always work. If the control animal is itself a generalised predator it can be a disaster (as in Australia, where the cane toad ate almost everything but the beetles it was aimed at).

Five thousand different insect predators have been used, worldwide, to control the aliens. After a great success in Australia with a moth that ate only prickly pears, conservationists have introduced a specialised predator of the scale insect, a ladybird that has beaten it elsewhere with no problems. In 2002 they let go 1365 beetles and are waiting – with a certain nervousness – to see if, on the Galápagos, they will continue to restrain themselves to a single prey.

Not long ago Darwin's islands seemed doomed in the face of fishing, tourism, population growth and alien invaders, but the joint efforts of the Ecuadorian government and the Charles Darwin Research Station have done a little to stem the tide. Not all the news is good, for the most generalised killer of all is still – in spite of the local seas being named a World Heritage Site – hunting sea squirts, hooking sharks and drowning sea lions in his nets. Only if we return to the narrow diet of the days before Darwin set out on his exotic journey is there much hope for the Galápagos, or for the rest of nature.

An Echoing Universe

I go to the Proms only now and then, although on most days I spend several hours with Radio 3 (often via the internet to get music while I work).

Concerts always strike me as rather an anachronism. That is not because of the High Victoriana of the Royal Albert Hall, but because music is so easily available in other formats. Nobody insists on reading manuscripts rather than books, so why the fuss about sharing a room with the band? Even worse, you cannot do much else useful at the same time.

Except, of course, daydream; and, at the Proms, the mushrooms are a great help (the giant fibreglass ones suspended from the roof of the Albert Hall, and not the traditional aids to hallucination). They are an attempt to improve the building's once-dreadful acoustics (certain seats, it was said soon after it opened in 1871, were better value than others because you could hear the concert twice: the original and its echo). The Hall's huge oval led, in parts of the auditorium, to more than a second's delay between the sound arriving direct from the stage and its reflection from the dome.

The fungus-like diffusers arrived in 1969 and, after much adjustment of their positions, have more or less cured the reverberation. Unfortunately, they also hide a magnificent fresco, soon to be restored, and must soon be shifted again – nowadays not by endless manipulation of the real things, but with computer models of the Hall and its virtual roof.

A great boon for architects, this new electronic age, for

now they can design a musical space at a keyboard rather than on a scaffold. It has done a lot for music too, from CDs to digital radio and television. I could have listened to my most recent Prom (Renée Fleming singing Mozart and Richard Strauss) on Radio 3, or watched it live on BBC2, with sound quality almost as good as my rather distant seat in the auditorium. But, oddly enough, echoes in the ether have come to haunt the virtual world, as once they did the Albert Hall. It will take a lot more than mushrooms to put them right.

One way to appreciate the problem is to tune an ordinary radio and TV, a digital version of each, and an internet connection, to *The One O'Clock News*. The pips sound, on each device considered alone, with what seems great accuracy. The Greenwich Time Signal, after all, uses the vibrations of a caesium atom, correct to one part in a billion, and American scientists have invented a mercury-based chronometer ten times better (they are hard at work on a clock that will lose a second in just three hundred million years).

Such precision might seem more than enough. But listen to all the time signals at once – digital and analog radio and television, together with the net – and the result is cacophony, with the noise of several sets of pips apparently echoing around the room.

The problem is that found in the Albert Hall: delays in transmission caused by a complex path between transmitter and receiver. In the good old days of Medium Wave, signals came straight from Crystal Palace (or its equivalent) to crystal set, with just a tiny hold-up caused by the wiring at each end and the finite speed of electronic beams through air. In digital radios or television, or on the internet, the affair is less straightforward. All kinds of compressors, decoders, distributed pathways, multiplexers and the like are involved, and each slows down the system – often in an unpredictable way.

As a result, the pips do not arrive on time. Even worse, for those with digital televisions, sound and vision can be retarded to a different extent so that while Renée Fleming's lips form the word 'Jubilate!' the sound might be 'Exsultate!' (The BBC, I should say, does much better than that in its own programmes.)

To synchronise the two in a broadcast is easier than to deal with the overall architectural problems of reverberations in the digital world. What is to be done to cure the echoes in the ether? The idea of giant mushrooms in each living room is not practical. What we need is a huge conference of everyone involved, from sound engineers to electronics experts – and where better than the Albert Hall? After all, its frieze says that it was 'Erected for the Advancement of the Arts and Sciences' but (apart from a talk by Stephen Hawking and the Imperial Bicycle Exhibition of 1883) science almost never gets a look in. And a scientific conference offers far more chances for daydreaming than does the most tedious Promenade Concert.

Drink, Desire and Performance

The opening paragraph of the US Army's *Mule Training Handbook* reads: 'First attract the animal's attention by striking it smartly over the head with a stout stick.'

The favourite cudgel when we move from mules to ourselves is, needless to say, alcohol. Its main effect is brutally simple, for it starves the brain of oxygen (although brain receptors are also involved). With the right dose one can, in theory, liberate the wellsprings of imagination from the brain's lower parts, where they are confined for most of the time because of the inhibitory effects of the drug on the command centres within the cortex. Five of the ten Americans who have won the Nobel Prize for literature – Sinclair Lewis, Eugene O'Neill, William Faulkner, Ernest Hemingway and John Steinbeck – were heavy drinkers (Pearl S. Buck, Saul Bellow, Isaac Bashevis Singer, Toni Morrison and Joseph Brodsky managed without the stuff). It worked for them, so why not for the rest of us?

Science has its eye on that expensive delusion. In an experiment that ranks with splitting the atom or unscrambling the genetic code, psychologists have set out to test whether drink aids creativity.

The task was easy enough. Give volunteers a variety of flowers and ask them to arrange the blooms into a series of 'aesthetically pleasing groups'. At the end of the session the judges awarded points on the basis of how different a subject's floral compositions were, one from another; in other words, how inventive the arranger had been. Alcohol came in because half the participants ('moderate to heavy social drinkers') had been given vodka and tonic. The other half had been served tonic alone.

In some ways drink worked, for those told that they had sipped vodka won a bouquet for creative work more often than did those who believed that their glass contained only tonic. There was, though, a twist to the tale. Half the alleged tonic drinkers had, without being informed, had their drink spiked with alcohol, and half the supposed vodka bibbers had drunk quinine-flavoured water.

In other words, the force that drives the flower arranger is not alcohol, but imagination. The drug has cruder effects, for in the same kind of experiment those who drank responded more slowly to a flashing light, whether or not they believed that their drink had been spiked. However, its creative effects, real as they might be, are in the mind as much as in the bottle.

Other drugs share the same ambiguous power. If nothing else, most people agree that a judicious joint makes music sound better and food taste divine. The brain's own cannabis-like chemical, anandamide, is involved in the passage of messages between nerve cells. Perhaps that too is an arbiter of good taste.

The experiment to test the idea is simple and obvious. Just like the alcoholic flower arrangers, half the subjects under test were given the narcotic or, with no warning, an innocent replacement ('by inhalation, ingestion or suppository') and asked to grade ('hedonically rate') how salty, sweet, sour or bitter a particular drink tasted. The result is, once again, simple. Those who believed that they had smoked dope even if they had not thought that food tasted better. Again, the beatific effects of the drug lie in the imagination.

On the wall of the Chelsea Hotel, in New York, is a plaque that records that from there, in 1953, Dylan Thomas 'sallied forth' to meet his end. The real affair was more sordid. His last words were 'Eighteen straight whiskies, I think that is the record!' and he died from what the hospital described as an insult to the brain, helped by an injudicious injection of morphine from a drunken doctor.

Drink killed William Faulkner and Dylan Thomas, as it has so many others. Thomas knew that his weakness had begun to inhibit, rather than liberate, his artistic powers. As he said a few weeks before he died: 'There is only one position for an artist and that is upright.'

Perhaps such experiments hold the key to Dylan's Dilemma. If the creative power of drink lies within the imagination, HM Customs and Excise should secretly arrange that half of all bottles of whisky (or vodka, or Chardonnay, or lager) be alcohol-free. There should be no hint of this on the label. In this way the nation would reduce its alcohol intake with almost no effect on its creative output. The same approach might be used with other vices – half of all butter could be margarine, or half of all cigarettes lack nicotine, without anyone noticing. Why, it could even work with television – one might get away with half of all comedy programmes without jokes. That experiment, at least, is well under way.

Will *Homo WalMartensis* Survive?

Make a list of the words used in this book, or any other, in order of occurrence. Some are common and some rare; but that banal fact hides a remarkable pattern. The British National Corpus – extracted from a hundred million words of written English – gives the big picture. 'The', for example, is number one; 'single' is 3804; and 'helix' lags at 9717.

Then draw a graph between the score of each word and the number of times it appears. A few words are very common, some have a middling number of appearances and hordes are almost never used. As a result, if the axes are set up in powers of ten – 10, 100, 1000 and so on – the line between the rank of a word and how often it turns up is straight.

That pattern is called Zipf's law, after its discoverer. It works for many apparently unrelated things: the size of communities (a few huge cities, a fair number of towns, innumerable villages) and the number of hits on internet sites (billions to Yahoo, millions to the BBC News site, tens to a home page). It extends to rivers (countless streams, plenty of Severns, one Amazon) – and smash your wife's favourite vase against the floor in an access of fury and the pieces, as you glue them together, come in sizes determined by that universal decree.

Zipf also rules the economy. In the United States over a million people work for WalMart – but the nation also has fifteen million enterprises that involve just one person. As a result, the distribution of company size over a millionfold range, from the self-employed to small building firms to gigantic combines, follows Mr Zipf's rule. Even India fits, for, to balance its hundreds of millions of peasant farmers, the country has the world's largest employer, Indian Railways, with a workforce of 1.6 million. So precise is Zipf that his law predicts that no enterprise will ever employ ten million people, for even the United States is just not big enough to support such a mega-corporation (the global economy, though, might allow just one).

The Zipf line has scarcely budged over two turbulent decades in the markets. Some firms crashed and some boomed; some were taken over and some split; but Zipf did not waver. The ecology of business has a fundamental rule: that – given the army of small companies – the emergence of a few vast firms is inevitable (indeed, the graph for the United States suggests a slight *shortage* of huge combines compared with expectation).

Zipf economics also makes a prediction about the fit between size and abundance in the living world. There are more mice than elephants and more sparrows than alba-trosses and, whatever group they belong to, large animals are rare. Just like elephants and mice, WalMart and Jo's Diner are at different ends of a remarkably straight line. As a result, the difference in the actual amount of elephant compared to that of mouse is quite small; there may be fewer of the latter, but it takes a lot of rodent to make one pachyderm. The rule works for most creatures, but the actual numbers involved vary among groups. Mice, rats, deer and other land mammals are, when species of the same weight are compared, a hundred times commoner than birds, and two hundred times more abundant than bats. (Britain contains a hundred and sixty thousand tons of wild mammal compared with thirteen thousand tons of bird, a useful statistic when next deciding whether to buy a pheasant in your local supermarket.)

Farms, and those who depend on them, smash the Zipfian guidelines. Ninety-five per cent of all the animal flesh in Britain is not buzzard, bat or badger, but sheep, pig and cow. Pheasants, too, are farmed in all but name, and are much more numerous than are wild birds of the same size. Such creatures are hundreds of times more common than they should be, while another familiar species, *Homo sapiens*, is ten thousand times more abundant than expected on body weight. In the unspoiled world of Nature the population of these islands would be around five thousand (about that of Marlborough, Wiltshire). Man flouts the rules to such a degree that the average Briton now meets more people in a day than his Stone Age predecessor would have seen in a lifetime.

It all turns on energy and efficiency. Birds and bats are less common than land mammals because they burn up more resources to stay in the air. Farm animals abound because man pours his own economic effort (much of it in the form of subsidy) into keeping them alive. For a sheep, it makes perfect evolutionary sense to live off the back of the taxpayer. But for how long sheep and their keepers can afford to defy the decrees of Mr Zipf, time will tell.

A Gordon's for Me

I was once interviewed by a journalist who reported, with a certain acerbity, that the cupboard in my office is full of empty gin bottles, with glasses scattered among the detritus on my desk.

True enough, but I am (I think) not an alcoholic. The glasses contained orange juice on a sultry day, and – like all academics – to relax, I drink only the finest Australian Chardonnay. The bottles themselves were part of a research project. They come from places as far apart as Argentina and Ethiopia. Each looks oddly familiar. A yellow label and bluish berries down the side, and an English title: Baron's in Ethiopia, Gunson's in Spain and elsewhere a vast diversity of equally unfamiliar names.

They are the raw material of a study of the evolution of advertising. Each bottle is an echo of that estimable English drink, Gordon's Gin. Go to the Third World and dozens of different versions are on the shelves, but in Britain only one. The Distillers' Company – which makes the stuff – has a negative view of people who copy its trademark. It fights off those who claim to be what they are not by making their brand name expensive to copy. As a result no cheats exist in the Distillers' native land, where the law makes deceit costly; a reminder that truth in advertising works only if the advertiser pays enough to ensure that his symbol means what it says. Without such investment, anyone can palm off a second-rate product as the real thing.

The pricey signal rule goes a long way. Even criminals follow it. Belgium, for a few months, had a much-feared

'Moustache Gang' of bank robbers. After a series of brutal robberies a strip of facial hair was enough to terrify cashiers into submission; but once there had been time for milder criminals to sprout the right cue, the intimidatory power of moustaches began to wane. In crime, as in business, the best signals are too expensive to imitate, which is why Japanese Yakuza gang members cut off their fingers rather than growing hair on the upper lip.

The same rules hold in biology. Take those familiar statements of sexual quality made by avid males. The peacock's tail, the toad's croak or the nightingale's song all cost a lot. Amorous (and noisy) male frogs are picked off as bats home in on their calls, and a bird may use half its energy budget in singing.

The best sexual signs, like the best gin, cannot be faked. For a toad to make a deep croak it has to be big; and, try though it might, a small toad cannot counterfeit a basso profundo enough to attract a female. The rule is more or less universal. The sex life of yeasts and fungi is real but dull, for, in their humble world, sex itself is no more than the fusion of two cells. A big healthy cell makes a desirable mate – and it needs to test the quality of its potential partners (of whom there may be hundreds).

It does so by forcing them to show their hands in a form of sexual poker. A cell in receptive mood produces a dose of chemical messenger. Anyone who hopes to mate with it must match, or beat, that card. The first player then steps up the signal and those who can summon up enough resources respond in kind. As the dance goes on each signal is countered by another more extravagant version until the weaker members of the team cannot afford to continue, and drop out. In the end, only the best – those who can afford a truthful but expensive statement of their qualities – are left to win the battle for attention and to collapse into what passes for their partner's arms. In much the same way a feeble male great tit with his black stomach band subtly enhanced with indelible ink does very well with the ladies, but he pays for it when he is beaten up by enraged males whose signals of rank mean what they say.

Now and again impostors manage to subvert such systems. The tropics have lots of tasty butterflies with conspicuous wings – which at first sight seems odd. In fact their patterns copy those of brightly coloured and poisonous species. They persuade potential predators, on the once-bitten-twice-shy principle, that the harmless mimics are noxious rather than nice. Quite often they pull off the fraud, but the trick works only as long as the frauds themselves remain rare in comparison with the noxious kinds.

As my gin project progresses I plan to test the parallels between the evolution of cheating in butterflies and in booze. One is already obvious. Many of the bottles in my cupboard are still full of a clear but sinister liquid. It took no time at all to find that, in spirit as much as in body, a cheap label means an inferior product.

At the Speed of Light

Speedy stuff, light, but how fast does it actually go? Once its rate seemed infinite, and Aristotle himself thought that light reached the Earth the instant it left the Sun, which was, he knew, a long way away. Galileo disagreed, and suggested a way to measure the crucial parameter. Two people, armed with a covered lantern each, should practise signalling. The first uncovers his light and the second, just next to him, does the same as soon as he sees the flash. Once they have established how long it takes for each observer to react, the trick is repeated with the two a mile or so apart. Any extra delay in response will then measure the speed of light. Galileo claims to have tried the idea but found no difference between the two attempts.

A clever idea, but – given light's real rate of travel – hopelessly unrealistic. The first practical test of that figure had to wait until a certain Hippolyte Fizeau did a most elegant experiment. His 1849 observation is a wonderful example of how science can progress sideways and of how logic is of no help in planning the next discovery.

From his house on the outskirts of Paris, M. Fizeau pointed a telescope at the window of his workshop on the hill of Montmartre, five miles away. As night fell he shone a beam of light from the house to a mirror in the workshop window. It was reflected back to where he stood. A system of gears spun, at variable speed, a wheel with seven hundred teeth. Each cog broke intermittently into the light beam as it set off on its ten-mile trip from suburb to city and back.

When the wheel turned slowly the light was reflected

with no hindrance, for it left the house and returned home in the same gap in the wheel, before the next cog had time to come into position and interrupt the journey back. As the machine speeded up, though, all of a sudden, at about twelve revolutions per second, the image of the lantern in the distant mirror disappeared, for the next cog had moved into place to block the reflected light by the time it had completed the traverse.

Measure the distance travelled, the space between the cogs and the wheel's rate of rotation and – *voilà!* – the impossibly fast, the speed of light, seven and half times around the world in a second, was measured.

The experiment was clever indeed, but what was the point? Did Fizeau improve the economy, or the quality of life? Certainly not, but by chance he was a keen photographer. A young astronomer wanted to track the movement of Venus across the sky. He added a light-sensitive plate to the machine and made a 'photographic revolver' designed to produce a series of images taken at great speed, one after the other.

Next a resourceful biologist used Fizeau's workshop to make a device (based, like the original, on a toothed wheel) for studying the movements of animals. Like its forerunner, his machine had a lantern, a lens, a mirror and a mechanism to interrupt a light beam. It was far superior to the earlier apparatus for serial photographs created by the British-born inventor Eadweard Muybridge (who shot his wife's lover with a *real* revolver). Muybridge had used a horse to break wires as it galloped and to fire off a line of cameras.

The final step in the saga was to reverse the process and to run the successive images through the cog-wheel apparatus until they merged to give the illusion of movement. Motion pictures had been invented.

From Fizeau's apparently useless experiment the film camera – and the industry it produced – was born. It took just twenty years. So much, then, for the futility of pure science, for Hippolyte Fizeau and his prototypical machine are a wonderful example of how good ideas, abstruse though they seem, can produce useful and unexpected results.

Albert Einstein, too, was among the least practical of men. He saw how the finite speed of light constrained the universe and how mass and energy were linked by Fizeau's constant. From his great insight came power stations and bombs, each valuable in its own way. The latest news is that light can, thanks to wormholes, tachyons and other mysterious processes, in theory sometimes go faster than light. Might the 'warp drive' so central to science fiction soon be here? To move at such speed would need all the mass in the universe to fire up the machine, so, for the time being, the price of a ticket around the post-Fizeau universe remains infinitely large.

On Fate and Genes

Walk into the surgery and within minutes the doctor can tell you roughly when you will die. Age? Sex? Weight? Blood pressure? Do you drink too much? A smoker? Any symptoms? Your income (useful, given the effects of wealth on health)? For most people that covers much more than half the information (your birthday adds more, for adult survival is worse for those born in the spring).

Medicine has done well in the battle against juvenile mortality; and if all deaths before the age of fifty were abolished by government decree, life expectancy in Britain would go up by little more than a year. It has not yet, alas, abolished death itself, which means that most people now die old, often of conditions about which rather little can be done.

Medical research has, as a result, become very interested in the diseases of the elderly. Many of them, from cancer to heart disease, have a strong inherited component. At first sight, then, the Biobank scheme, set up in 2003 with a start-up from the Medical Research Council of forty-five million pounds, is a step in the right direction. Never mind the squawks from those whose cash was cut to pay for it (their successors will squawk louder, for this is just the downpayment on a thirty-year project), genetics will soon become the key to a healthy life.

The scheme is simple, but grand. Take half a million people in their fifties and sixties, persuade them to give a blood sample and to bare their souls about their past and present lifestyles. Then set up a series of molecular gridlines along the DNA

and follow their lives and deaths in the hope of tracking down
the genes behind the inevitable illnesses of old age.

Well, maybe. The first problem is that the extent to which
inheritance is expected to influence life expectancy and the
miseries of the years is, for good evolutionary reasons, limited. The old adage 'If you would live long, choose your
parents well' is quite correct – but has little to do with DNA.
Surrey parents and children live longer than those in
Scunthorpe or Somalia, but because of the differences in
the way they live, and not in their biochemistry. Family
studies show only about a quarter of the variation in adult
lifespan is even indirectly attributable to genes (which is less
than the proportion for height or for IQ).

All this makes biological sense. Natural selection does not
work well on the aged for they have already had their children and passed on their DNA. As a result, and just as in an
old banger, decay is inevitable, but hits separate and unpredictable bits of the machinery. There are no genes for old
age and no reason to expect much consistency in what sections of the genome kill off each one of us. Already dozens
of genes – different genes in different patients – are impli-
cated in Alzheimer's disease, and for cancer the tally is even
higher. Why spend vast sums to search for more?

Other difficulties with Biobank are more practical.
Some subjects will be recruited through doctors, which
cuts out people who do not have one and brings in those
who are already ill. The doctors themselves have to be
computer-literate, which means a bias towards richer practices and patients. Although Biobank has tried to enlist a
wider audience, even a small skew of this kind can be
important. Thus, almost all the health differences between
American blacks and whites, once blamed to some degree

on DNA, arise instead from the stark and simple fact that blacks are poor.

Even worse, the Biobank survey looks resolutely backwards. It is bad enough asking people what they did yesterday, let alone forty years ago. I've never smoked – but how much beer did I drink when I was a student? (Too much, for sure.) And what did I eat? (The Scottish Suicide Diet plus a few apples.) Sexual partners? (Nostalgia, mitigated by hyperbole.) Seventy-five thousand Britons are in medical cohorts whose lives have been followed from birth, some to the age of sixty. Why not check their DNA instead, or that of people already diagnosed with disease, rather than this arbitrary group?

Other problems are more subtle. The hot stuff nowadays is gene expression – what is switched on and in what tissue, when things go wrong? – and a blood sample says nothing about cells hard at work in heart or brain. Why spend millions on high-tech DNA sequences when the crucial information on health comes only from notes taken on a brief visit to a GP's surgery plus an interview with a nurse, or from a death certificate filled in by a busy doctor?

And, if you are tempted to use a curved dagger of oriental design to reason with Biobank's architects, bear in mind that the police can apply for access to the database.

A Twitch on the Worldwide Web

Snails are not much of a winner at dinner parties. My only molluscan discourse arises from the inevitable question: 'Do you eat them?' and the answer is 'No!', which kills that social interaction.

Think how much easier it is for those who work on spiders. The animals are a goldmine of conversational gambits. Raw spider tastes of potato mixed with lettuce. The standard Brazilian recipe goes: 'First dig out your tarantula. Hold the head down with a stick and remove the fangs, toast over a fire to burn off the hairs, serve sliced with chillies.' The largest species, the Goliath, is at ten inches across just big enough to fill a dinner plate. Many Americans believe that the average citizen eats four live spiders in his sleep each year, but no scientist agrees with them.

Spiders can be good for you. The Rev. Dr Thomas Mouffet dosed his daughter Patience (of tuffet fame) with spiders to cure her ills. ('The running of the eyes is stopped with the dung and urine of a House Spider dropt in with Oyl of Roses . . . There is in England a great lady still living who will not leave off eating them.') Nowadays a drug based on spider venom is used to treat heart attacks. Those who plan to give a tarantula as a pet should remember that a spider is for life and not just for Christmas, for the animal can live for twenty years in captivity.

As he compliments the host on the wine, the arachnologist may wish to tell him how Californian vineyards have straw houses to protect spiders over the winter, ready to attack pests in the spring. Wine snobs even introduce boxes of the creatures into their cellars to add an authentic ambience. Should the social occasion be a wedding breakfast, the talk could turn to the fact that in 1876 the Chinese ambassador presented Queen Victoria with a gown made of spiders' webs. Since then silk genes have been cloned and can produce the ideal material for wedding dresses. Hindus, indeed, scatter the animals instead of confetti at weddings.

Those who plan to attend a matrimonial event in the American South might be warned that the poisonous Black Widow is common in outside privies and attacks the first part of the body to be lowered on to its web. Women are nine times more liable to be frightened of spiders than are men; but men are, for the most fundamental reasons, more susceptible to the spider's kiss. Many soldiers in the first Gulf War believed that the so-called Camel Spider would eat large chunks of human flesh, but the animal is in fact not a spider at all and is harmless.

Few dining rooms, if searched hard enough in a post-prandial game of Hunt the Spider, fail to turn up a specimen. The creatures are everywhere. A fall of gossamer in 1940 was (briefly) thought to be a chemical warfare attack. By 'ballooning' on silk, spiders can travel a thousand miles from land. On Everest they go higher than any other creature – to twenty-two thousand feet. Thirty-five thousand different kinds are known (and only about twenty pose any real threat to humans), but not all, alas, are safe. Some – the No-Eyed Big-Eyed Wolf Spider of the USA springs at once to mind – are on the World List of Endangered Species.

If the event is a dinner-dance, spiders can be a helpful ice-breaker. In medieval Italy the cure for a bite was to dance a tarantella, for hours or even days. The species involved is in fact quite mild. As the performers were allowed to behave as lewdly as they wished, perhaps the event was more of an excuse to relieve rural tedium than a specific against arachnid poison. To cure the sickness, a victim was buried up to his neck in dung and surrounded by seven female dancers. If he laughed, he would survive; if not, death was certain.

Should the audience show signs of boredom they can be reminded that some basic questions remain unanswered. The National House Spider Survey of Great Britain has tried to resolve them. Do the animals prefer detached houses or semis? Soon, we should know.

Finally, over brandy and cigars, one could relate how spiders illustrate the philosophy of science in a classic experiment on arachnid hearing. Shout at one and it runs away; pull off its legs and it does not. This proves, of course, that its ears are in its legs. I used to quote this to undergraduates to show the limitations of the scientific method; but all I succeeded in doing was to produce a generation of students strangely ill-informed about spiders.

On the Rocks

The bottles of fizzy water served up by certain American airlines carry a cautionary notice: 'Small crystals may appear in this liquid in cold conditions. These are of no danger to health and will disappear upon warming'; a statement reminiscent of the bag of peanuts that is supposed to have said: 'This product may contain nuts.'

Such admonitions sound absurd (although, given the real dangers of peanut allergy, the second is not as foolish as it seems). The famous crystals are, needless to say, ice; and those who take their airborne H_2O to dilute a different liquid would be better served by a health warning on the whisky miniature (as indeed happens on some airlines).

But ice is less simple than it seems. Cool water slowly in a domestic freezer and it may stay liquid – until the container is given a sharp tap and the whole lot at once goes hard. To move from one phase to another – from disorder to order, liquid to solid, water to ice – takes energy. Once a few molecules have been forced together (a sudden blow helps) the great shift of state happens at once, as it spreads from a single point.

A speck of dust (banned, no doubt, from bottles) also helps to persuade reluctant water molecules to get together as it acts as a local nucleus to which they can attach themselves. That reduces the endless vibration intrinsic to the liquid state, allows them to overcome their natural disinclination to join and sparks off the big freeze.

Space is a great cathedral – a large, cold, empty and sometimes rather grimy place. Stardust now and again forms solar systems (our own included) but most of the time it stays as dust. As on Earth, it acts as a centre of attraction for the other chemicals that float in minute quantities through the heavens. Without such crucial places in which molecules can get together in the almost vacant universe, they would almost never come into contact; but upon the tiny particles – just as in a bottle of chilled water – the molecules of the void can meet their neighbours.

As a result, the interplanetary grains are covered with ice and solid carbon dioxide, together with some less predictable elements of the stellar cocktail. Certain corners of the universe serve up copious quantities of alcohol. One single star in the Aquila galaxy, ten thousand light years from home, has a cloud of ethyl alcohol around it big enough to provide a bottle of the finest Johnny Walker for every person on Earth every day for the next five thousand billion years.

Far above the heads of the determined tipplers of business class, life itself may have started as an airborne, albeit dusty, whisky and soda. In the icy cold of a Dutch laboratory, a quick blast of ultraviolet light of the sort given out by the Sun on to an artificial version of the icy and alcoholic stardust (plus, alas, a dash of ammonia, an ingredient absent from any beverage I have ever tried) has caused some unexpected substances to appear.

They are amino acids, the building blocks of life. Some are identical to those found in our own bodies. If polarised ultraviolet, whose waves vibrate in only one plane, is used in the experiment, the new molecules are twisted to one side – as is the case for our own amino acids (but not for those made in a chemical laboratory, which give a mixture of two mirror-image forms). Some of the dustiest parts of the sky do emit polarised UV – which may explain why the building blocks of life on Earth lean so much in one direction (the left, as it happens).

Lab experiments that try to mimic the chemistry of space are bound to involve plenty of guesswork and the claim that life began in outer space is almost impossible to test. Even so, NASA's *Stardust* probe is flying through great clouds of tiny particles at the edge of the solar system and collecting them in a special net. It has already raced past a comet and collected the fine fragments that stream in its wake. When the vehicle gets back to Earth in 2006, after a trip of two hundred million miles, we may find out just what they are made of and whether bränd-new versions of our body chemicals are indeed floating through the void in a great interstellar distillery. The thought deserves contemplation over another glass of something cold as the flecks of dust left after the Big Bang rain gently down outside.

No Bells or Whistles

Most American television is garbage: cartoons, cop shows, news channels that would be at home in Hoxha's Albania, all endlessly interrupted by screaming ads.

There are, though, certain exceptions. One programme is so laid-back as to be almost comatose, for the Research Channel features on-camera lectures from academics. It was funded by a group of research universities and is now available, free of charge, on satellite and cable, twenty-four hours a day, to more than twenty million households. Switch it off! comes the cry; but in fact the material is oddly irresistible. Each speaker is an enthusiast who takes the job with great seriousness – and they have no option, as they face student audiences who do not hesitate to ask hard questions.

The choice of subjects is best described as eclectic. A typical evening might include shows on small molecules that interfere with cell division; on how to prevent giant telescope mirrors from becoming distorted when their glass (itself a super-cooled liquid) flows like water; and on a new contraceptive gel that kills off the human immunodeficiency virus.

Just to prove that there may be life beyond science, the channel also has talks about autobiography and adoption, but for most of the time pumps out a stream of disconnected facts from the vast world of chemistry, physics and biology. It can be funny – an astronomer discusses how his observatory was thrown into a panic when the telescope went black and something dreadful seemed to have gone wrong (they had forgotten to open the dome). Sometimes it verges on the odd – the gel man assumed the shifty expression shared by everyone who discusses sex on TV. Whatever its quirks, and however unexpected its topics, the energy and expertise of those involved shine through. All human life is there: new anti-cancer drug found in a bacterium unique to Easter Island; how mobile phones interfere with radio astronomy; and more than one would wish to know about the acidity of body fluids.

It succeeds because it is interesting, and because it is honest. Nobody fakes fascination. Instead they are fascinated and it shows, for, after all, they are talking about their life's work. The presenters manage with no music, no drama, no sob stories, no doom-laden futures, no ethical bleats, no desperate attempts at relevance; just the straight story, straight to camera. In fact the Research Channel does what the world's academics do every day as its presenters tell the audience what they are up to and where they hope to go next (although in the real world the fact that the lecturer sets the exam does help). Some of the stuff is perhaps too technical (who really wants to watch half an hour on spinal reconstruction of kyphoscoliosis, healthily grown potatoes or women in history?) but most is accessible to anyone with an interest in the world around them.

Compare that with a lot of the television science in Britain. A great deal is first rate, but the subject suffers more and more from the nervous tic of giving constant reminders of how marvellous life is rather than allowing life to speak for itself. *Horizon* has had a dirty bomb in London, the brain hard-wired for religion, women and Viagra, and DNA and the Hanratty murder. Why not the straight stuff on nuclear physics, brain chemistry, pharmacology and gene diversity, rather than science as soap opera?

A simpler approach could pay off, in much the same way

as it has for cookery programmes. Find some great communicators and let them talk to a camera as they do to their
undergraduates. The raw material is there. One of the few
useful things to emerge from the blame culture that per-
vades British universities is a huge body of student
assessments of their lecturers. The average is pretty good,
but among the broad mass are scattered a number of
Grade-A talents unknown outside their own walls. Why not
ask each university to identify its top six, and give them a
screen test?

A pastiche of the lectures in the unused universe of
hidden genius, like those compilations of sadistic clips from
Japanese game shows, would give a new insight into the
world of science. Any competent producer could generate
fifty minutes of hallucinatory strangeness from the Research
Channel alone, flipping from the edge of the universe to the
biology of sewage works. To mix that with the similar, but
more professional, output from our own Open University
would say a lot about the two cultures on either side of the
Atlantic – and think how much cheaper it would be than yet
another hour on houses and gardens.

The Case of the Brown Dog

In the summer of 1903 a distinguished scientist at University College London handed over a large cheque to his employer. Dr William Bayliss, of the Department of Physiology, had just won a libel case. The Hon. Stephen Coleridge, an antivivisectionist, had accused him of deliberate cruelty to a brown dog which had been operated on in front of students.

The story was a fabrication and Bayliss sued – and got two thousand pounds (a sum equivalent to almost a hundred thousand today). The anti-science league was as expert in deluding the public then as it is now and twice the amount was raised from animal lovers by subscription. The excess went towards a statue of the brown dog itself. It was later destroyed by UCL medical students, in the worst riots to hit London until those sparked off by the Poll Tax.

Things have not changed. A substitute bronze pooch lurks in Battersea Park, the bench upon which the crucial experiment was done still exists and the antivivisection movement carries on with its lies about science. Activists are still trying to persuade the public, as they did in 1903, that animal experiments are not needed.

A century ago the claim was quite untrue, for the famous canine had contributed to a great breakthrough in biology. Bayliss had found that when food comes into contact with a dog's small intestine, a chemical messenger travels in the blood to the pancreas and persuades it to pour out its juices. Secretin, as he called it, was the first hormone to be discovered (and he invented that word, too).

Secretin is just one of many protein-based hormones that control the body's energy budget. Its prime job is modest, as when food enters the intestine it turns on an alkaline tap which quenches the acids of the stomach and allows digestion to proceed. Two of its many relatives, ghrelin (from the Greek 'to grow') and (a less digestible title) PYY 3-36, have a more arduous task, for they set the scales of hunger and satiety and decide when to demand dinner and when to push away the plate.

Ghrelin is made in the stomach and tells the brain, 'Fill me up.' Its opponent senses the presence of a bellyful (bulky foodstuffs are best, for junk food is digested too quickly) and sends a 'Time to stop' message from its home in the intestine. The two hormones interact with others to control the energy budget. Insulin is also involved in the equation, and far more human lives have been saved by that substance than animals sacrificed in research upon it.

The concentration of ghrelin shoots up just before a meal and drops afterwards. The success of the gastric – and drastic – surgery sometimes done on the morbidly obese may as a result come not from a simple reduction in the size of the stomach but from a drop in its ability to make the dinnerbell hormone. A complete bypass can cut down by three-quarters the amount of ghrelin produced (and a pill to interfere with the famous protein might remove the need for operations altogether). Ghrelin reduces the amount of fat the body burns, which is why it can be hard to lose weight once it has been gained, for the obese get as hungry as the rest of us, even if many pounds of only too solid flesh seem to be available to fulfil their needs.

To understand those internal messengers may help to control today's plague of obesity. Biologists are hard at work to find out what lies behind the epidemic. They use their most sophisticated tools – cells in culture, computer models, genetic tests and modified versions of the protein. Surely, in this post-genomic world, animal experiments are no longer needed?

Not at all: animals are still central to medical research, and many of the studies done today are no more than high-tech versions of the labours of Dr Bayliss. A search of the hundred most recent scientific papers to include the word 'ghrelin' in their title shows that ten involved laboratory tests on cells or tissues, thirty-nine used human subjects – and fifty-one were experiments on animals. Without their help, we would be nowhere.

One of the fixed untruths of the anti-science movement is that such work is old-fashioned and can be abandoned. Its relative importance has indeed gone down, for the actual number of procedures has not changed since the 1950s, while the science of biology has exploded. Even so, the idea that we can ban such experiments at no scientific cost is quite wrong. No doubt the same will be true a century from today when – with luck – medicine will understand obesity. And no doubt the activists will still be assuring anyone who will listen that animal experiments are a waste of time.

Brown Dogs Revivisected

'Now, what I want is, Facts . . . Plant nothing else, and root out everything else. You can only form the minds of reasoning animals upon Facts: nothing else will ever be of service to them.'

Thus Mr Gradgrind in Dickens' *Hard Times* and thus the scientific method: a foundation of solid and agreed fact upon which further grains of information may tentatively be piled. Not, on the other hand, the logic of many of its opponents, as fact can always be denied and flighty fancy allowed to run amok.

The controversy about animal research has been much discussed in the newspapers. Some points are reasonable and open to debate – that the inspection system may be flawed, or that we should take more account of animals as sentient beings (although I question the view that the work should be done on genetically damaged children instead). Most of its opponents, though, use an essentially stupid mode of argument: that if you disapprove of something, it cannot be true.

Europeans for Medical Advancement, for example, insist that 'small genetic variations between species create profound biological differences that preclude extrapolation from one species to another', while Animal Aid feels that 'hundreds of millions of animals have been killed in laboratories in Britain alone. Not only have their deaths been utterly in vain, but they have also caused the deaths of millions of human beings.'

Such claims are simple nonsense. Tamoxifen, a drug used

against breast cancer, was developed from studies of hormonal change and cancer in rats. An anti-cancer drug tested on mice had turned out to kill the parasite responsible for the tropical disease kala-azar – and this led at once to a treatment. Thalidomide – once the cause of many birth defects – was not tested on pregnant animals first. Now all drugs are so checked, and an anti-acne medicine found to cause problems in mice was banned, which meant another tragedy avoided.

Perhaps the greatest triumph of all – the control of diabetes with insulin, which saved millions of lives – emerged from work on dogs. It has been supplemented by experimental transplants of the damaged endocrine cells in pigs and monkeys. The research has turned to the chemical cues on the surface of mouse cells that lead to the fatal attack on the pancreas by the immune system. It has not yet produced a treatment and may never do so, but without such work there can be no hope of progress. All this has involved the sacrifice of many animals but not, I think, 'utterly in vain'.

The brainless parroting of fancy in the hope that it will turn into fact is a common tactic in the strange world of science-denial. Creationists use it all the time. They claim that carbon dating is a fraud, that there are no intermediates in the fossil record or that the cell is too complex to evolve. Rubbish; but rubbish endlessly repeated can convert itself in the public mind into uncertainty, and then to truth.

Scientists are well advised to steer clear of the proponents of devoted stupidity. In 1870 Alfred Russel Wallace – famous for his part in evolutionary theory – broke that law, and paid the price. John Hampden, a convinced flat-Earther (is there any other kind?), had bet five hundred pounds that nobody could prove that the surface of a body of water was anything but flat and Wallace, short of cash, took him up on it. The experiment was simple: a long canal near Bedford with three bridges; a weight lowered to the same height above the water from each one; and a sighting along the line which showed that the central weight appeared to be above those at each end – a result, needless to say, of the curvature of the Earth.

From his efforts Wallace gained nothing. He was pursued by True Believers for the next decade. A hint of the tenor of the debate comes from a letter to Wallace's wife: 'Madam – If your infernal thief of a husband is brought home some day on a hurdle, with every bone in his head smashed to pulp, you will know the reason.'

Wallace was no conventional scientist. He opposed smallpox vaccination (and was wrong) but he did so by finding flaws in the sums done by supporters and not by a flat assertion that it did not work. He was against animal experiments, too. In his paper 'Is Nature Cruel? The Purpose and Limitations of Pain', published after the brown dog case, he wrote: 'I myself am thankful to believe that even the highest animals below ourselves do not feel so acutely as we do; but that fact does not in any way remove my fundamental disgust at vivisection as being brutalising and immoral.'

I disagree with his view, but respect it. Most biologists feel that, with proper controls, the gain from such experiments outweighs the pain they may inflict. Wallace argued from an ethical stance and not by ignoring scientific facts. In some ways he succeeded, for the rules are far stricter than they were in his day.

Would that today's antivivisection movement had half his honesty – but an ability to lie to yourself is the first requirement when your case rests only on denial of the truth.

Take Your Partner

I spent most of one summer on a diet of frozen vegetables – the peas were OK but the spinach was better. Usually I never touch anything taken from a deep freeze as it reminds me too much of those lab freezers full (as mine was the last time it was cleaned out) of obscure little bottles labelled 'Elephant Serum?' or 'Could Be Sperm'.

There was a scientific reason for this change of diet. The frozen packages were to keep the slugs cool. I was hard at work in one of my favourite haunts for snail work, the Val d'Aran in the Spanish Pyrenees, once unspoiled but about to become the route of yet another Europe-subsidised motorway.

Slugs have an uncanny ability to transform themselves into a loathsome soup when they get hot. As the refrigerator in my battered old field vehicle broke down long ago, the only way to avoid slime bouillon was to keep frozen food plus slugs in an insulated box.

Collecting snails in Spain attracts no attention. Spaniards eat plenty of them, and in the Val d'Aran I have been mortified to see some of our numbered specimens appear as lunch. Near by are caves filled with millions of discarded shells thousands of years old. After all, why bother to chase a mammoth when just as much protein can be obtained from a creature unable to run away? Slugs are different. I have heard rumour of an Italian dish, slugs baked in milk; and a slug held in the mouth was once a cure for tuberculosis. But no one seems to like them much, and the locals were bemused to see us at work.

Nevertheless, limacology, as the science is called, has its fascinations. As Hitler probably did not say, slugs are degenerate snails. Each group has evolved from an ancestor with a shell. The animals gave up their external armour because of the expense of making it. In evolution nobody gets a free lunch, and the price for a naked snail is to be doomed to the night and the fog. No slug can come out in the heat of the day, and the driest parts of the world (the Val d'Aran excluded) are free of them altogether.

Some slugs have given up another talent. Most molluscs are hermaphrodites and contain both male and female sex organs. In spite of the temptation that this must offer, in general they lead lives of rectitude. Boy-girl meets girl-boy and nature takes its course. For certain slugs, though, the allure of availability is too much. They go in for self-fertilisation, the closest thing to incest, and what Woody Allen may have meant when he talked of 'sex with someone you really love'. In time their descendants evolve into strains of what are, in effect, identical twins. A few years ago we began to map the distribution of such strains. The logic was simple: to find out what persuades some slugs to abandon sex might hint at what got the whole business started in the first place. We have not solved that problem but some clear trends do exist.

In Europe southern slugs are a lot sexier than those from the north. Big black slugs, from Scotland to the Arctic Circle, are all one genetically identical strain. They fertilise themselves. Those from the laxer cultures of southern England and Spain go in for sex instead. What is more, it seems that the abstemious northerners are being overwhelmed by a wave of sexuality spreading from the permissive south.

All this fits the Calvinistic stereotype common to Scots and Scandinavians, but why should molluscs show the same pattern? Climate may have something to do with it. If – as in the steamy tropics – one's main enemies are the other animals who are out to get you, it is risky to put all one's biological eggs in one basket, as asexual creatures do. But if the only opponent is a predictably dreary northern climate, it may be safe to evolve into a single strain that can cope with the weather. One of the unexpected benefits of climate change (or perhaps it is all those food imports needed to indulge in the Mediterranean diet) may be an increase in the sexiness of northern populations, of slugs at least.

That is why we were in the Pyrenees. Perhaps, up in the cool mountains, even southern slugs would take up self-fertilisation – or perhaps we might find some relict groups that had retained this austere way of life since the Ice Age.

For two weeks we struggled up mountain roads (and even walked, a pastime to which no field biologist will normally admit) in a hunt for the high-altitude self-fertilising slug. We were unlucky. Even at seven thousand feet it seems that sex prevails. And that is a typical result for a field trip, or any research project, where a good idea turns out not to work after all. But in case you think this was a complete waste of taxpayers' money, don't worry: I paid for the spinach.

The World Turned Upside-Down

Black, sooty black; and dank to match, with little more than a set of tawdry and tentative lights to illuminate the filth of London's streets. December is the cruellest month, with the shortest of days and the most dismal of seasons. A Scrooge-like attitude, maybe; but shared by many. Some people can afford a break in a place where the sun shines and the day is twelve hours long; but most of us have to cope with cold, darkness and gloom.

It was not always thus. Camden once had a time of endless sunlit days – and in December, to boot. At Christmas half a billion years ago the whole of Britain was on holiday abroad.

To be more accurate, foreign parts had a trip to the British Isles, because this country (or where it sits today) was much closer to the equator. Its change of venue was due not to continental drift but to something much more dramatic. Suddenly, some geologists claim, the whole world flipped and shifted the axis around which it spun. Before the reversal, what is now the equator was near the poles. Today's poles, in turn, basked in subtropical sunlight.

The first evidence lay in the remnants of great glaciations near the modern equator. The signs of their activities as they scoured the Earth are clear, but how was it possible to freeze the tropics and spare the poles? Was the entire planet turned into a giant snowball, with every continent covered in ice a kilometre thick? Life could never have survived such a universal chill.

A hint of the truth lies in ancient records. The Chinese

accounts of lunar eclipses, moments when the shadow of the Earth falls across its surface, are old indeed. More than a thousand years ago an astronomer noted that: 'In the first rod of the third watch, the Moon was seen in the clouds above the direction *ping* . . . above the direction *ting* the eclipse was total . . . it was restored at the end of the direction *wei.*' Such records, together with more recent observations, cover hundreds of eclipses over thousands of years, but when astronomers return to the early figures the theoretical course worked out for each eclipse is (or so it seems) hundreds of miles away from where the ancients actually saw them.

Something in the equation must have changed. We know that days are getting longer by about one and a half thousandths of a second each century as our planet is slowed by the Moon's drag. However, the Moon's mass ought to be enough to slow the spin down by an extra half a millisecond in each hundred years. The apparent anomalies found in the ancient eclipses show that something else has speeded us up, and shortened our days, since those times.

The end of the last ice age, a geological instant ago, is to blame. As the ice caps have melted slowly over the past ten thousand years, the load on the poles has gradually decreased. As a result, our planet is less flattened at its northern and southern ends than before and so spins slightly faster – which explains why Chinese eclipses do not fit and why December days are shorter than they ought to be.

A much older Armageddon of ice had a far larger effect on the planet's shape. Nowadays our planet is flatter by fifteen miles at the poles than at the equator. In those ancient glacial days, hundreds of millions of years ago, it was squashed flatter still by polar ice many miles thick. To make things worse, the drifting continents were bunched together near the poles and their mass distorted the world to make it even more like an orange than it is today.

This change in shape was enough, the physicists calculate, to alter the balance of the spinning top on which we pass our lives. It forced the Earth to flip. As a result, the tropics moved to the poles and the poles returned the compliment. Before the shift, the world turned on a different pivot, today's Arctic and Antarctic basked in tropical heat and Britain was south of where the Mediterranean now sits. As a result, the ancient glaciations whose traces still remain in Africa and central America took place, not near the equator but – just like those of today – at the Earth's northern and southern ends.

The final proof lies on the Moon. Its orbit is at an angle to where planetary theory says it should be – but only if the Earth has always had its present tilt. Assume that it has capsized, and the Moon's orbit makes sense. An ancient flip may explain the globe's glacial history, but is it any consolation to learn that half a billion years ago our planet was even more depressed than it is today?

Dr Strangelove, I Presume

Jonathan Miller's right arm is detached from the rest of his body, in political terms at least. He tells the tale of looking down his republican nose at an enthusiastic crowd as the Queen drove by, only to find that his own hand had decided, with a will of its own, to wave frantically to the Royal Personage, against orders from Central Command to stop.

As so eloquently seen in Kubrick's most famous film, sufferers from anarchic hand syndrome must battle with their unruly member as it tries to give Nazi salutes, steal food from a neighbour's plate or even strangle its owner. In a few cases the errant hand reverses what the other one has just done – it closes a newly opened tap or unbuttons a shirt after it has been put on. Some patients even deny that the appendage belongs to them.

The condition may arise when an accident or a stroke damages a part of the brain's frontal lobe called the supplementary motor area. Whatever the cause, to have an anarchic hand can be traumatic. Sometimes it can be treated. One man's hand insisted on pulling at his pyjama collar to wake him up, but could do so no longer when he wore an oven glove in bed. Drugs may also help.

Like many such disorders, the unmanageable hand gives an insight into how the brain works. It also has a ghastly fascination of the kind much relied upon by popularisers of science. They defend their prurient approach on the grounds that the public needs all the science it can get, even if it takes a certain voyeurism to persuade them of that fact.

The problem with popular science, of anarchic hands or

anything else, is that it does not know when to stop. Scientists themselves are in part to blame. They abuse words in an attempt to make themselves more interesting than they are. Some turn to sociology – the science of humans – to describe the world of animals. The term 'rape' is used, absurdly, to describe the sexual habits of ducks (birds famous for their moral responsibility), whose females may drown when they suffer the attentions of a male. Political correctness can swing the other way. One research paper in three on topics such as the effects of testosterone on behaviour uses 'gender' rather than 'sex' in its title, much as that brings to mind the image of a sea elephant in a frock.

The supplementary motor area can, when damaged, free hands from conscious control. As a result, it has been sold as the locus of free will. Such damage, some say, unleashes the devil within, who, given a chance, does bizarre and anarchic things. But is this biology or speculation? Is there – for hands or any other body part – a measurable entity called free will, or is it a useful but unscientific concept, like 'good' and 'evil'?

To lose control of one's limbs is a problem hard to ignore. 'Free will', in contrast, is an ambiguous notion whose home is in philosophy rather than science. Is it one phenomenon, or many? Can we identify that magical stuff except when it has gone astray? Do chimps, or mice, have any? Why not call the supplementary motor area the centre for creativity or for religion instead – for, after all, is it possible to paint or pray if your right hand has a mind of its own?

Philosophers see free will as a state of inner uncertainty that allows each person to choose between different courses of action. To scientists that idea is mere illusion. None of us chooses to breathe in and out, and nobody can hold their breath for ten minutes even though they might wish to do so. Triumphant though their will might seem, nerve signals from the brain force them to defy it and to inhale. In fact every action is preceded by such neural activity and the mechanisms involved are no more complicated than are those behind breathing. Science, some say, simply does not need the vague notions central to philosophy.

The law, on the other hand, does. What if an anarchic hand were to strangle someone? Who would be to blame, the limb or its owner? Lawyers recognise freedom of will by its absence, but that absence may have many causes. People who hear voices that instruct them to kill are seen as less culpable than those who kill to steal a car. Schizophrenics with their inner voices (and genes lie behind that awful disease) are treated as ill rather than evil. Drug abusers with the same problem are not – which is not consistent, for nobody doubts the existence of inborn and quite involuntary differences in the ability to deal with narcotics.

Where does science stop and subjectivity begin? Philosophers know, but scientists, as usual, are not quite so certain.

Thumbs Down for Politicians

Every election is an experiment in statistics. On most occasions one party wins; but, under the British system of first past the post, often with a majority that does not much reflect the number of votes cast. We boast about our democracy, but the uncomfortable fact remains that almost all our post-war governments took power on a minority basis.

The dilemma gets worse as more candidates enter the fray. In the election which marked the re-election of the first Blair government in 2001, a shift to the Lib Dems of 12 per cent would have allowed Labour to win with a third of the votes and more than half the seats, while the Lib Dems themselves, with almost as many votes, would have obtained just one seat in seven. As it happens, the Lib Dems did rather worse than that in terms of votes and won only one seat in twelve.

Statisticians spent years in attempts to build a better political mousetrap. Sad to say, their mathematical cousins have now proved that the perfect electoral system in an election with more than two candidates does not exist. Even so, dozens of different designs have been tried. Proportional representation, an election determined by the fraction of the country as a whole that votes for a particular candidate, is popular – but how to do it? By division of the entire national ballot, or by transfer, with those with fewest first-choice votes kicked out one by one until just a single candidate is left? Both have problems: unstable coalitions of tiny parties for the first and voter bafflement for the second.

Those who understand the problem have come up with an answer. The Mathematical Association of America and the American Statistical Association now elect their committees by a new method called approval voting. It could not be simpler – vote for as many candidates as you like, without ranking them in order. In the British political context, if you love the Greens, like the Trotskyites and have a certain sympathy for the UK Independence Party, tick all three boxes; and leave the others blank. Each vote carries the same weight. Nobody's vote is wasted and, computer simulations show, the result is the best available reflection of the consensus view.

In the first George W. Bush election, Ralph Nader won it for the Republicans as his three million votes were lost to Gore. Under the approval voting system the result would have been different. Most Nader enthusiasts were liberals and would have added Gore to their ballot, while none of them could stand Bush. In the absence of fraud (always a challenge in American elections) Gore would have won the presidency, as he did the crude voter count. The result of the 2004 replay under approval voting would have been less predictable, but perhaps the uneasy coalition of Christian fundamentalists and those alarmed by September 11 would have fallen if some of the latter had made a gesture towards the Democrats as well.

Not all elections succumb to the mathematicians' approach. In politics, as a rule, each election has more voters than it does candidates. For book prizes the opposite is true. A hundred authors of prodigious talent put their efforts before the jaded eyes of half a dozen of the moderately great and not unreasonably good. The electorate – often half a dozen or fewer – is so small that, at least on the first round, each voter may decide that he prefers a different winner. In turn, that can lead to undignified wrangles and real inconsistency.

Thus, the 2001 Samuel Johnson Prize for non-fiction shortlisted several books involving thick-skinned creatures that sift through trash. Most were biographies, but one (Richard Fortey's *Trilobite!*) was not. As a science book it had, most unusually, made it to a general non-fiction short-list. However, it failed to attain even that modest eminence in the Science Book Prize for the same year – proof of how unpredictable the decisions made by a small electorate may sometimes be.

We need a fairer method. The Soviet Union had the answer: a system in which voters crossed off candidates of whom they actively disapproved. The idea seems odd, even anti-democratic, but to mathematicians it is just the same as approval voting, a tick against the ones you like, even if it feels quite different to voters (and, no doubt, to those who get elected). When it comes to books, I can't stand imagined historical narrative, or footnotes, or poetry, or crime fiction, or violence, and many other things – which at once cuts out most potential prize-winners and makes the choice far easier.

Yes, disapproval voting, that's the British way. So many people dislike fake biblical grins, bespectacled and sinister Welshmen or chubby but affable Scots that the Green Party, the Socialist Workers or UKIP would be bound to win the next election. Then Parliament could come up with another method for the next time round.

Some Like It Cold

The coldest I have ever been was in a thin sleeping bag in a British midsummer. Unfortunately, I was on the other side of the world, and of the seasons, at the time. The camp was in a desert in South Australia and I was taking part in a futile search for the Hairy-Nosed Wombat, then thought to be on the edge of extinction (we found just one, in a cage outside a pub). When the June sun went down after the Austral winter day, the temperature and our spirits each took a plunge.

African primates like ourselves find it hard to deal with cold. Give people from the equator or the poles the choice and they set the central heating at exactly the same level. We northerners (and South Australians) cope by taking the tropics with us in the form of duvets, double glazing, central heating and holidays in sunny places. All this is expensive – as those who vacation in the Seychelles find out – and most Britons spend more than half their income in the battle to keep frigidity at bay.

Cold is dangerous, with an increase in deaths due to heart disease and stroke of 10 per cent for each drop in temperature of three degrees. As a result, the number of strokes in winter is nearly twice that on a hot summer's day; and cold living rooms are enough to explain the winter peak in the death rate in Britain among the poor compared with the rich.

Most creatures do not have the chance to make their own climate. Instead they are forced to suffer what nature throws at them. Plenty of animals are cold for most of the time –

those in the Alps and the Arctic, of course; but also the denizens of the deep sea, which covers most of the planet and whose temperature hovers just above zero. Evolution has done a lot to help. For them, life slows down. Their enzyme molecules bend more easily and need less energy to chew up the raw material that comes in food than do those of their effete relatives from warmer places. Certain people, too, have evolved to face low temperatures. Eskimos, with their short arms and legs, are better at keeping the heat in than are the elegant and lanky Kenyans.

To add a soluble chemical to any fluid drops the freezing point (which is why salt helps on icy roads) and many creatures protect themselves with antifreeze of various kinds. Some face a bigger problem, for the thermometer may take a sudden plunge to far below zero and they turn into solid ice. Many mountain and Arctic animals undergo that petrifying experience but recover. A certain Scandinavian beetle manages fifty degrees below zero (much colder than a domestic freezer) and comes back to life when warmed up. Canadian frogs can turn into solid blocks and thaw out with little harm, for their cells are kept safe with a special antifreeze. Those hardy beasts have other tricks, for they dry out when the mercury falls – which cuts down ice formation – and make blood that is very ready to clot and plug any leaks when the thaw at last arrives.

Many of these adaptable creatures chill out with ethylene glycol, the stuff put into car radiators in winter, while strawberries (themselves susceptible to ground frost) use glucose to do the job, which is why they taste so sweet. If you use the right frog, Pete and Dud's frog and peach pudding would make a tasty dessert, for the glacial Canadian amphibian uses the same sugar to protect its cells (and how it manages a hundredfold increase in blood glucose is of interest to doctors specialising in diabetes). The antifreeze proteins of Antarctic fish are so good at their job that they have been engineered into bacteria to make soft ice cream that can get really cold before it sets hard. True Greens might prefer the frog pudding.

Certain animals convert a fifth of their mass into alcohols to save themselves from freezing. Not even the most determined human drinker can match them (although his toad-like hangover does come in part from water loss and his post-binge heart attack from the increased tendency of the blood to clot that accompanies it).

People do not make antifreeze, however cold they get. On the other hand, those kept in icy rooms for a week do become more tolerant to the chill. Their levels of thyroid hormone increase, they shiver less, their hands and feet are warmer and they do not raise the temperature of the body core as a defence. Indeed, Australian Aboriginals cope without shivering on desert nights far colder than those that a European can handle.

In other words, to be sure of feeling warm and comfortable in a chill winter, wherever it may be, lay off the drink and writhe in the snow for an hour a day.

A Ghost in the Ladies' College

Three-dimensional kind of place, Cheltenham. Rolling hills, modestly soaring architecture, and even the horses are obliged to leap over fences to get to the finishing post. It has some intellectual heights, too, with its Festivals of Literature, Music, Jazz, Folk and – a latecomer on the scene – Science. For several June days its streets are alive with people keen to discuss why diamonds are for ever, whether mankind will survive the century and what's new in the twin sciences of Beer and of Old Age (other recent popular talks have included 'Exploding Custard' and 'Blown-up Biology').

For a place in which life is lived on so many levels, it seems odd that the event has never celebrated the town's leading scientific son, the Oxford-educated mathematician and eccentric Charles Howard Hinton, born in 1853. In 1880 he published in that seminal journal the *Cheltenham Ladies' College Magazine* (he was the ladies' maths teacher at the time) his great work entitled 'What Is the Fourth Dimension?' There he introduced the idea that, beyond the familiar up–down, left–right and near–far world in which we spend our days, there exist other dimensions to reality, which can only be inferred, rather than seen.

That concept, then unfathomable, has become commonplace. The Victorians were astounded by Muybridge's serial photographs of galloping horses, for the images made visible a fourth dimension – time – and many people simply could not understand what they meant. Nowadays we are more sophisticated. Anyone can draw a graph of the relation between height and weight, or, with some nifty pen work, between height, weight and blood pressure; but although they cannot draw it, most people can see the relationship between those three measures considered together and the chance of a heart attack. Statistics tries to make the complex simple by reducing the number of dimensions in which we describe the world.

The notion gets into physics, too, for Einstein's relativity depends on an additional dimension, with gravity emerging as a consequence of curved space-time. Since then the idea has run wild. Until not long ago the most popular model of the universe involved a single point, which, at the Big Bang, gained three spatial dimensions plus a fourth, called time. Such a notion is itself pretty hard to grasp, but things have got worse. String theory, which attempts to unite all the forces of physics – gravity, electromagnetism and the rest – needs lots of dimensions; there were once, it claims, ten of space and one of time (although six were carelessly misplaced as the universe was formed). The latest perversion, brane theory, has it that the universe has been multidimensional all along. It started when two rippling parallel membranes collided after a long period in five-dimensional space. The energy released became electrons, photons and so on; and the points where the membranes touched each marked the beginnings of galaxies. This also has an odd resonance of Hinton, for his fourth dimension, too, was wavy (which, for him, had something to do with consciousness).

The Science of Beer is easier than that sort of stuff, which seems to many perhaps undereducated minds to be as much mysticism as physics. Hinton's own paper was republished as a pamphlet entitled 'Ghosts Explained', which imagined a sort of superhuman who could annihilate or create any object in our familiar world simply by taking it to, or moving it from, the fourth dimension. A century before Rubik, its author invented a series of eighty-one different-coloured cubes that could be manipulated to illustrate his ideas. They were marketed as tools to allow those who mastered them to view the world of the dead. Hinton taught himself, or so he claimed, to be the first man to think in four dimensions. That equally flaky character Salvador Dalí picked up the idea and used it in his famous painting of a floating Christ crucified, *Corpus Hypercubus,* in which the Cross is based on one of Hinton's cubic constructions.

Charles Howard Hinton died in 1907, appropriately enough, for he was a bigamist, while proposing a toast to 'female philosophers'. Cheltenham has missed its chance to celebrate his science, but the town could make amends in some future Literature Festival. Hinton himself wrote mystical fiction (pain, in his books, was a message from beyond), H. G. Wells was influenced by his work in *The Time Machine* and some claim to see Hinton's mark in the writings of D. H. Lawrence.

Hinton was a man of wide talents. He fled to the United States to avoid imprisonment for his marital crimes. There he designed a gunpowder-powered baseball machine that could be adjusted for pitch and curve. After some nasty accidents the batters were afraid to face it; but it was the forerunner of the computer games that Cheltenham, so far, has kept out of its august halls.

Sexual Chemistry

Scientists have a mania for defining the obvious. One early description of man was 'An upright, bipedal vertebrate, without feathers' (which at least cuts out the chickens). Some of his attributes, though, seem beyond definition. It was a sociologist, not a scientist, who designated love as 'the cognitive affective state characterised by intrusive and obsessive fantasising concerning reciprocity of amorant feeling by the object of amorance'.

But even love – the only four-letter word not yet in regular use in polite society – has begun to yield to the scientific method. If 'reciprocity of amorant feeling' means anything, it refers to the tendency of two individuals to select each other from the crowd and to stick together. The life of the prairie vole shows how science can help understand emotion. Conservatives could adopt this virtuous creature as a mascot, for the noble vole is obsessively monogamous. A male sticks with his mate through thick and thin and any intruder audacious enough to step into the marital home is driven off.

The machinery behind such commitment is remarkably simple. It depends on a hormone for faithfulness – a molecule just nine amino acids long called arginine-vasopressin, perhaps better referred to as 'infatuin' or 'devotin' ('love-in' doesn't sound quite right). Within minutes of mating with his chosen female for the first time, a male prairie vole becomes protective. His loss of virginity leads to a sudden jump in the amount of the devotion hormone, and given a choice of two females, he runs at once to his spouse.

So powerful is the hormone that a single dose makes even a virgin male stick with the female he shares a cage with, although he has never been allowed to mate with her. On the other hand, an injection of a chemical which blocks devotin's action will persuade any male prairie vole to run off with the next female to come along.

The discoverers of the magical molecule point out rather sniffily that 'the extent to which any single peptide subserves any aspect of social bonding in humans remains entirely speculative'. A vole does not know what love is (or, if it does, we do not know how to ask it). Although humans also have the hormone, and men secrete more of it when aroused, we have no idea whether human sexual chemistry is as simple as that of the prairie vole. However, the reciprocal amorance of many mammals is certainly driven by the same kind of machinery.

The sex life of the laboratory mouse is not like that of the prairie vole, or our own, for a female faced with a new male abandons her old partner at once. She will even end a pregnancy to be ready for her paramour. She does not need to see him – the heady fragrance of urine is enough to turn her head. In a remarkable experiment on mouse morals, animals engineered to contain the prairie-vole version of the cell receptor that responds to its special hormone change their behaviour and become paragons of monogamy.

For mice not engineered towards purity, the choice of mate is in part inborn. Specialised genes, situated among those of the immune system, encode a set of individual scents that allow the female to tell one male from another. She hunts for novelty and, above all, avoids sex with a mouse who shares her own scent and might therefore be a relative.

Among the faithful voles of the Ohio prairies live the

Hutterites, a religious group who migrated from Switzerland. Like many of those who flee from bigotry, they despise outsiders and marry within their own group. As a result, they are highly inbred.

Chemical signals seem to be involved in Ohio amorance, for married Hutterites are less likely to be similar to one another for certain genes in the immune system than are pairs who are just friends. Quite without realising it, most Hutterites – and perhaps most people – fall for someone with a set of identity cues different from their own. What is more, they try hard to avoid a partner whose genes are too much like those of their own mother. The Hutterite mother (or perhaps anyone's mother) is to be avoided as a role model when choosing a wife. Just how the system works, no one understands; but there may be a hint of scent in the air.

The first social engineer to bottle devotin may change the habits of the most irresponsible father. On the other hand, the real money is to be made by patenting the antidote.

A Pinch of Salt

When it comes to statistics, there are lies, damned lies and the tobacco industry. One recent lie about smoking is so dreadful as to make one gasp and stretch one's eyes. It leads to some unpalatable truths into how medicine uses – and abuses – statistics.

Most studies of whether people are damaged by other people's smoke are too small to be reliable. Put all the information together, though, in what is called a meta-analysis, and the effect is clear. In particular, children whose parents smoke are, on average, less healthy than those not exposed to the poison. In the most persuasive report on that work, the graph shows not only the average health of the two groups but, as usual, the error bars – a sign of how much variability is present in each set of figures. The error bar around the passive smokers' score just overlaps the average health score for children from families in which nobody smokes.

To the tobacco lobby, with its bizarre logic, that proves that in some families smoking is actually *good* for children! Their elementary mistake (surely it cannot be deliberate) will, no doubt, help persuade addicted parents that their family is one of the lucky ones, and to carry on puffing.

To prove that this is a rant about statistics rather than tobacco – what about salt? We are all aware that a salty diet causes high blood pressure and many people try to cut down with nasty low-sodium stuff. Nobody, after all, wants hypertension and it seems a sensible precaution.

Well, good try but, as smokers (and salt-makers) say, no

cigar. A meta-analysis of more than a hundred trials of salt reduction shows little overall effect. Sure enough, some studies do show that salt is bad for you (just as some show that tobacco is good for babies). Given the laws of chance, they should. After all, to spin ten thousand pennies ten times each means that one will, more often than not, come up all heads. That coin is not magical, for it succeeds only because of the accidents of sampling.

The trouble is that we – and those who cause, or cure, cancer – are fatally susceptible to Magic Penny Syndrome. The problem is sometimes referred to as the Bing Crosby Effect, after his song 'Accentuate the Positive, Eliminate the Negative'. Our attention is drawn to positive results and we ignore those that do not fit our ideas. The best-designed salt study was done in Scotland in the 1980s and found that intake had no effect on blood pressure. A follow-up in 1998 showed that there was no effect on death rate, either. The result had little impact, the low-salt lobby is still hard at work and the statistical argument goes on.

The first real use of statistics in medicine was a mere half a century ago, in a 'randomised trial' (in which neither doctor nor patient is told who has received a drug until the experiment is over). In 1948 the new Medical Research Council carried out the first such test, of the effects of sulphonamide on tuberculosis. It gave a strongly positive result and the substance was put on the market.

Before then, medicine depended on tradition, anecdote and reputation. If Sir Lancelot Spratt thought a drug worked, or that a leech could do the job, people listened, for no better reason than that he was famous. Nowadays a toss of the coin is the norm and who gets what in a clinical trial is determined by chance, doubly blinded. Enforced statistical honesty has saved thousands of lives and the figures show that it pays to do the job properly. A meta-analysis of dozens of trials of drugs against cancer and heart disease shows that trials not based on randomisation are five times more optimistic about a treatment than are those that include the safeguard.

Even haphazardness has its hazards and the beautiful truths of mathematics must yield to the brute facts of the real world. Many medical trials, even randomised ones, are so ill-designed as to have no chance of a sensible result. Nearly all the two thousand or so tests of the effects of drugs on schizophrenia involved too few patients to have any reasonable hope of finding an effect, two-thirds were deemed useless by the experts and just twenty were judged to be satisfactory.

Even so, and in spite of the problems with randomised trials, it is still a good idea to take what the tobacco lobby says with a grain of salt.

Losing its Savour

Lot's wife looked back and turned to a column of salt. Here I follow her lead. The previous piece claimed that there was no credible evidence that salt was bad for health. It was based on a paper in *Science*, first among equals of the world's scientific journals. Things could not have been worse had I been outed as a Sodomite (an inhabitant of a mountain of sodium chloride in the Negev) for the piece produced a chorus of disapproval from medics who insisted that I was absolutely wrong.

Salt raises passions as much as it does blood pressure. A string of editorials about the substance have the word 'controversy' in the title. Although several studies have indeed failed to find an effect on health, on balance the evidence seems clear: cut down if you can. The controversy comes from the fact that only in some places do the figures fit.

Salt was first a luxury, became a preservative and grew into an addiction. Our intake is ten times more than it was during most of our past (although it has dropped since British food was the salt of the Earth, before refrigerators, when brine kept it from rotting). The Romans were so hooked on the white powder that they added it to wine because so much was present in their food that otherwise their drinks seemed insipid. The same trick can be played on chimpanzees and children – start them off on a high-sodium diet and they will starve rather than eat unsalted food.

So powerful are the myths about that addictive additive that saucy wives once gave it to their salaried partners at bedtime because of its salutary effects on their salacious performance. Its talent, like that of other drugs, comes from its ability to ask for more of itself. Salty food suppresses the action of taste receptors so that a normal diet seems bland. (St Matthew knew all about it: 'Ye are the salt of the Earth; but if the salt have lost his savour, wherefore shall it be salted?') Anyone can wean themself off the fatal chemical, but most people see no need to try. In Britain the average consumption is about ten grams a day and many doctors would like to see the figure halved.

Half the population has high blood pressure by the age of sixty, which is bad news because it much increases the risk of stroke and heart attack. Other addictions – tobacco, alcohol and the Pill – are in part to blame; but quite a bit may be due to the hidden narcotic, sodium chloride. It is easy enough to cut down, for even a smaller hole in the salt shaker makes a difference. As most dietary salt is in processed food such as sausages and tinned soup, cook your own meals; eat more fruit and use your common sense.

But should we bother, given the claimed ambiguity in the figures? The Salt Manufacturers' Association hated the Sid the Slug campaign, which asked people to cut down, but even they admit that those with high blood pressure should take care – but should the same be true for everyone? Why should the rest of us worry?

Certain studies – like the Scottish one – show little effect on health, but others do. Japan has a greater intake of salt in the north of the country, matched by a higher incidence of stroke. In Portugal one village was persuaded to halve its intake and, within a year, saw a drop in average blood pressure. In Britain, by contrast, a decade ago the amount of salt in bread was secretly and slowly reduced by a tenth but this was not matched by a change in health.

Part of the story is a question of inheritance. Some people – those of African origin most of all – are sensitive to salt because of their genes. The same is true for many other drugs. The sodium story may become a microcosm of the future of medicine. Once we were just a population, but now we are individuals, each with his or her unique capacity to deal – or not – with particular stresses, be they salt, food or drugs. When we know exactly who we are, then the phrase 'choose your poison' will have real significance; and for some of us salt will indeed lose its savour.

Nobody recommends that people smoke more, or take up cocaine because they can handle it. On balance, the trade-off between bland food and a stroke has led me to give up salt. My receptors have reduced themselves to such a degree that I even taste the stuff in white bread. I still have no idea what my blood pressure might be.

Pentameters on Physics

A poet's account of a famous discovery:

> *When Newton saw an apple fall, he found*
> *In that slight startle from his contemplation –*
> *'Tis said (for I'll not answer above ground*
> *For any sage's creed or calculation)*
> *A mode of proving that the Earth turn'd round*
> *In a most natural whorl, called 'gravitation';*
> *And this is the sole mortal who could grapple*
> *Since Adam, with a fall or with an apple.*

To celebrate Einstein Year – the centennial of the great man's most creative twelve months, in 1905 – the British Association for the Advancement of Science launched a poetry competition on the themes of time, space and energy. Would that Newtonian verse gain laurels from the British Ass? It should; for it rhymes, it has metre – and the author is Lord Byron, which ought to help. The lines come from his epic satire *Don Juan*. Their rules are those of *ottava rima*, a witty form invented by Italian sonneteers. The composition scheme is *abababcc* – three matching rhymes, followed by a fourth and different one – while each line is an iambic pentameter: five pairs of syllables arranged as short-long, short-long, short-long, short-long, short-long. *Ottava rima* has a catchy pattern, with plenty of energy to drive it forward, but to find the rhymes and keep the rhythm is not easy – although Byron manages more than a thousand such verses in his poem.

As Robert Frost put it, poetry without rules is like tennis without a net. What makes *Don Juan* great is, surely, its rigid structure. The content can be unexpected, almost random ('Let us have wine and women, mirth and laughter, / Sermons and soda-water the day after') but the form is a framework upon which to hang words and images. Physics after Einstein is much the same: an uneasy stand-off between order and chaos.

Why do we see the poetic order as adding an extra dimension to the mere chaos of prose, to transform it into something finer? It turns on how we handle language, which more and more appears to be a combination of talents rather than just one. Teachers have long asked children, as they learn to speak or to write, to sit in a circle and to clap in time to a nursery rhyme. Their 'Hickory Dickory Dock' approach works because infants can identify rhythms and rhymes several months before they learn to read. They find it easier to remember a story when it is presented as poetry; and they chant verses, often with incoherent words, very young. Indeed, those best at rhymes and rhythms tend to speak earlier than those who are no good at them.

Different sections of the brain light up when we read poetry rather than prose; and when a child recites nursery ditties parts of that organ not associated with regular speech come into play. Some even claim the presence of a specific 'poetry module' deep within the skull (if so, Lord Byron must have had an enormous one).

That magical module sees patterns in words – which is what dyslexics cannot do, for they find it hard to translate the rhythms of spoken language to the shapes of sentences on a page. A failure to identify rhymes early in life correlates with later dyslexia, and such children do not succeed at poetic metre, either, for they are less able to pick up the beats in an electronic tone than are those with a normal ability to read.

There is science in poetry, and poetry in science. The Byrons of that latter art are, without doubt, to be found in physics. Our nation's experts, alas, will have no time to take up the British Association's challenge, for they, like all their fellows, are forced instead to spend their imaginative efforts on that prosaic exercise in Byronic deception, the Research Assessment Exercise. The poet himself would be safe from the government inspectors, for he continued to produce until he died at Missolonghi; but the hero of relativity might be in trouble, for his best work was done forty years before his retirement. How, I wonder, would Einstein fare in today's market place? Only poesy can tell the tale.

ODE TO A ONCE SUCCESSFUL, BUT NOW SUPERFLUOUS, PHYSICIST

Now comes the time, so bureaucrats proclaim,
To cease your work; from science avert your eyes,
To strive instead to seek – and then to blame –
The undeserving, and your kin chastise.
Pluck from the stars a modicum of fame
As once again (to no just man's surprise)
Luck favours those who play the rulers' game.
You, like our Lord, the televisual Winston,
Have not published: get thee out of Princeton!

Byron! thou shouldst be living at this hour.

Wandering Minstrels of Science

There was in Meiji Japan a Floating World – a society of aristocrats and courtesans who met in exotic places to exchange ideas (among other things), only to drift apart and come together somewhere else years later.

Apart from the aristocrats, the courtesans and the location, international scientific meetings are much like that. The great Congress of Genetics attracts thousands, myself included. At each event I find myself surrounded by people I have seen at five-year intervals since the first I went to, in Berkeley in 1973. To my eyes, none of us seems to have changed at all but, like an ageing Japanese roué, I have the uneasy sensation of a new and vigorous generation breathing down my neck.

Such events generate an odd sense of lost time. All around are the glazed eyes of colleagues trapped in conversations they thought they had escaped from in Toronto several years before. Lots of the eyes are bloodshot as well. The Professional Conferrers are here in force. They fly in for a day and then back to New York or Sydney after a talk, a schmooze with the funders and a round of plaudits from the peasantry. With so much of it about, jet lag becomes an infectious disease and the locals reel out of the conference hall as shattered as if they, too, had just stepped off a twelve-hour flight.

Another way to divide up the delegates is into those who wear their name badges in the street and those who do not. Wherever they find themselves, most of the Brits belong to the latter group. They share an air not of scientists but of someone who has come to look at the drains. However distinguished they may be, they succeed remarkably well: Professor Paul Nurse FRS, a Nobel Prize winner, seen outside any conference, looks far more like a window cleaner than my own window cleaner does. The Japanese, Russians and Germans, in contrast, stride the streets with suits, ties and identities proudly displayed.

A courtesan of the conference circuit faces certain perils. The worst is the need to produce a new discovery each time. There was once a Russian scientist, an expert on beavers. The last twenty or so of his hundreds of publications were each called 'More about Beavers'; which is accurate, but somewhat dull. At scientific congresses many papers suffer from the same problem. Genetics goes for strong elements of 'More about DNA Sequences', while gene therapy – the insertion of DNA into people whose genes have been damaged – is well into its 'too much more about' (or, to be more honest, 'not much more about') phase.

Science in the floating world is often mixed in with commerce. The machinery and the chemicals for research are expensive, with plenty of stalls anxious to promote the latest models. I was once at an event at which Sir Alec Jeffreys gave a talk about DNA fingerprinting. It was a dramatic account of how genetics has been used to trap the guilty and free the innocent. But all of a sudden came a brutal reminder that science can be misused. One American asked a long and complex question about its legal implications that was fielded by the speaker; for the questioner was acting for the prosecution and was desperate to be able to say that his views had the approval of the man who invented the technique.

I missed the last International Congress of Genetics,

which was held in Melbourne in 2003. My excuse was jet lag, but the meeting had, to judge from its programme, 'Rather Less about Gene Therapy' than for several years. The technology had improved – no more DNA sniffed up in oil droplets to treat cystic fibrosis, but modified viruses injected instead – and it seems that a few children with a rare immune disorder had even been cured by the injection of the appropriate piece of the double helix. However, there was much discussion of the disturbing news that some patients had developed cancer, perhaps because the gene had got into the wrong place.

The great ship of conferences still sails endlessly across the globe. The next great Genetical meeting will be in Berlin in 2008. No doubt the egos will be as enlarged and the eyes as glazed as they were in Birmingham. Nowadays the latest buzzwords have to do with individual differences in the response to drugs, in the genetics of common diseases and the degree of inherited variation in human behaviour. Perhaps, by the time of the next congress, we will also have learned 'Enough about Gene Therapy' for it to be useful.

Out of Time

Has anyone noticed that something odd has happened to Big Ben? As a regular viewer of ITV's *News at Ten*, I often check my watch against it; and, on almost every occasion, when its face says 10.00 my own trusty timepiece (accurate to a couple of seconds a week) says 10.02. A quick call to the Speaking Clock shows that the Nation's Timekeeper, not my watch, is at fault: Big Ben is running slow.

Britain's time is out of joint and there must be a scientific reason why. My guess is that the pendulum is somehow involved. Newton used the pattern of swing of a heavy weight to support his argument that the mutual attraction of two bodies – the Sun and the Earth, or the Earth and a clock – depends on their mass and the distance between them. For a large ball like the Earth, and a small one like a pendulum bob, the mathematics makes it convenient to measure the distance from the centre of each.

Newton was told by a navigator that his pendulum clock, which kept time in London, lost two minutes a day at the equator. With a brilliant intellectual leap, the great astronomer deduced from this that the pendulum swung slower in the tropics because it was further from the centre of the Earth than it was in England. In other words, the globe was not a perfect sphere, but was flattened at the poles.

From the change in the rate at which the mass beat back and forth in different places, Newton worked out that the radius of the world was one part in 230 less at the poles than at the equator. Today's estimate (based on satellites, not clockwork) is one part in 298 (which is around fifteen miles), which means that Newton was pretty close. One guaranteed way to lose weight is, as a result, to move to a hot country, where you are further from the Earth's centre of gravity than in New York or in London.

A clock might lose time in such a place for quite a different reason. The pendulum warms up, lengthens and slows down. The copy of Big Ben that sits in Buenos Aires may suffer from that problem – but can it really be true that ITN introduces its programme with a fake tropical clock face that runs slower than its London equivalent? It does seem a little improbable.

There could, though, be another explanation for the slowing of the nation's time. The Earth is not, as a physicist might hope, a homogeneous sphere. Some of its rocks are dense, others less so. This produces local changes in the force of gravity and hence alters the clocks. A Cornish timekeeper ticks at a higher rate than one in London because its pendulum is attracted by a huge mass of granite rather than the lightweight clay upon which the capital sits. Such gravitational anomalies, as they are called, can be spectacular. The ocean is two hundred feet higher in some places than in others because the dense rock below attracts billions of gallons of seawater. In the Pacific such watery bulges mark a line of massive seamounts that stretch westwards beneath its surface.

The anomalies have a practical value. Prospectors use them to search for ore, for if they find a sudden blip in the gravitational field, there may be a mass of heavy metal down there, ready to be extracted. One such lump sits below central Portugal, although nobody has yet found out what it may contain. Areas where gravity is feeble, on the other hand, mark light sedimentary rocks and the possibility of oil. Paris is one such lightweight place, which gives hope to its people that they may at last be able to break away from their addiction to nuclear power.

And that leads to another explanation for the National Time Anomaly: the Earth's mass around Big Ben has decreased, slowing the beat of its clock. There can be only one reason: our rulers have furtively shipped out the country's gold reserves. Is this a hint of a scoop to come?

Those opposed to Europe may also notice that, because of the Earth's rotation, when the time in London is ten o'clock it is several minutes later in Brussels. Perhaps the ITV *News* has been set to the clocks in the European Parliament as part of the slow erosion of our national sovereignty. Or could it be that the schedulers, anxious to increase the number of viewers for the ads in the expensive pre-*News* slot, have furtively delayed its start by two minutes? The idea is just too absurd to be taken seriously.

Gall, Wormwood and Van Gogh

Camden Town on a warm spring evening, with its multitude of bars, is a potent reminder of the joys and miseries of alcohol, from the sippers of Chardonnay to the swiggers of cider. Hidden in a backstreet just off the main party drag is a dingy Georgian house with a white plaque saying that Verlaine and Rimbaud stayed there during their sojourn in London in 1872. Each was a member of that great society of addicts of *la fée verte* (green fairy liquid, to provide an unromantic translation) – absinthe.

Absinthe contains a poison, wormwood, an essence extracted from a small shrub common around the Mediterranean and found also in sage and in certain cypress trees. Its effects were considered so noxious that it was banned by the French government in 1915 in the interests of the war effort. Science has found out how this magic narcotic works.

Thujone, the active ingredient, is a set of rings of ten carbon atoms, whose structure looks rather like that of the active agent of mint (and is not dissimilar to tetra-hydro-cannabinol, the prime ingredient of cannabis). The chemical is available over the net as a herbal tonic claimed to reduce anxiety, but to drink enough of it leads to hallu-cinations, convulsions and paralysis. Because wormwood is a painkiller that also kills flies, intestinal worms and malaria parasites the green fairy was, nevertheless, once given to troops as a health tonic. To the surrealists (and to the Frenchmen who sank more than thirty million litres of the stuff in 1910) it gave a double intoxication, alcohol plus something more. No wonder the state banned it.

The drink was brought to France from Switzerland in 1797 by a Monsieur Henry Pernod; and the rituals of taking it (most of which were aimed at hiding the bitter taste – the Greek *apsinthion*, undrinkable) involved a spoonful of sugar, upon which the absinthe was dripped. Then it was set on fire and plunged into water to give a cloudy liquid. Today's Pernod, a pastis (or pastiche) of the original, with several of its herbs, looks and tastes much the same but is wormwood-free (although the real stuff has been put back on the market by a firm in the Czech Republic).

Thujone attaches itself to the receptors of a neuro-transmitter called gamma-amino-butyric acid. These come in various flavours, but each is a gatekeeper in the brain, a controller of the release of messages between nerve cells. Such structures are the targets of many psychoactive drugs (such as alcohol itself) and may also be involved in epilepsy, schizophrenia, Alzheimer's disease and, perhaps, addictive and thrill-seeking behaviour in general. Anti-depressants such as Valium attach themselves to GABA receptors, as do other chemicals such as the 'loco weed' that sometimes kills cattle. Even the insecticide dieldrin finds its home there.

Absinthe has joined the list, for insects resistant to diel-drin also survive a solid dose of thujone, and a radioactive label shows that the thujone molecules make straight for the famous receptors themselves. They avoid, though, the sites to which tetra-hydro-cannabinol attaches itself so that absinthe's hallucinatory effects, whatever they may be, are not related to cannabis; indeed, thujone excites and alerts the brain rather than, like alcohol or cannabis, depressing it.

That, then, is why Alfred Jarry careered around Paris on a bicycle with his face painted green and why Van Gogh, with the help of the viridian leprechaun, cut off his ear.

Degas' famous Absinthe Drinker, depressed though she looks (and of whom it was said that 'Absinthe makes the tart grow fonder'), had in fact a set of buzzing nerve cells in her head. Oscar Wilde had a dramatic view of the stuff: 'After the first glass you see things as you wish they were. After the second you see things as they are not. Finally, you see things as they really are; and that is the most horrible thing in the whole world.'

Alas for enthusiasts of decadence the sums do not add up. In fact the amount of thujone per glass was so small that any drinker would fall off his chair for more traditional reasons long before the hallucinations set in. To suffer from thujone toxicity he would have to drink fifty green fairies in a row – which, given that the potion is twice as strong as whisky, might have certain side-effects of its own.

However, the cheap absinthe in the Belle Epoque contained plenty of other chemicals. Analysis of the few bottles to have survived shows traces of methyl alcohol (the stuff in meths), of copper salts and aniline green to improve its colour and of antimony chloride to make it cloudy when water was added.

All those, no doubt, did plenty of harm; but the sad fact remains that the big killer in absinthe is also found in Chardonnay and cider. The great decadents would have felt at home in today's Camden: Verlaine died of drink at the age of fifty-two, and Rimbaud at thirty-seven.

Measure for Measure

Some people measure out their lives in coffee spoons, but there are many other ways to count up the days. Science casts its ruler over every aspect of human existence and can put figures on both its pleasures and its pains. The real problem is to translate the units used into some universal language. We have the international henry, the gry, the therblig, the defective year, magnetic reluctance, the Lawson criterion and the Prandtl number; none of which impinges upon our everyday lives.

Other measures do. They begin at birth. A newborn's crib might be decorated with an Apgar score of the state of its health one minute after birth, and the brightness of its lucky star assessed in apostilbs or blondels (*pi* times the luminance per steradian). As the child grows, its IQ (the relationship of mental age to chronological age) is tested again and again.

Adolescence is tracked on the Tanner scale, sexual attraction on the Kinsey equivalent. When it comes to wedding bells, a happy couple can identify the mathematical formula used by the bell-ringers, from Plain Bob Minimus with a mere 24 ($4 \times 3 \times 2 \times 1$) combinations of notes, to Stedman Doubles with 5040. The candles at the ceremony may burn at the rate of one furlong-fortnight (which is almost exactly a centimetre a minute).

At the reception some people are content to open a rehoboam (half a dozen bottles) of champagne, but why not go for a salmanazar, twice as large? Cheap wine comes in a demijohn (a bottle with two handles – after the medieval French Dame Jeanne), but how big such a vessel might be is not at all clear. One could order a Russian garnet (3.28 litres; 15 garnets being 4 vedro, 40 schtoffs, 64 boutylki or 400 charki) or a Turkish oka (0.61 litre), but caution is needed as a Bulgarian oka is a mere half of that quantity.

Not everything is so simple. A wine gallon (3.7853 litres) is quite different from a gallon of wine (4.5459 litres). The former is the measure of volume taken by the Americans on independence, and explains why there is less gasoline to a dollar than British visitors think. One should remember, though, that an American dry pint is one sixth larger than a wet one. The glug is, alas, not a reliable substitute, for it is a measure of mass (the amount accelerated by one centimetre per second per second by a force of one gram).

And what glasses to use for a toast? A wide choice apart from the standard wineglass (28.4 centilitres) is available. Perhaps a Roman acetabulum (0.066 litres) or, if that seems a little small, a French chopine (0.465 litres) or its Scottish equivalent, three times larger.

Then comes the festive meal. A nide of pheasants, a sute of wildfowl or a brace of partridges might do. The birds may have been shot with a twelve-bore (a 0.729-inch barrel), but the size of the jovial peasant who fired at them varies in a baffling way from country to country, at least if traditional measures based on open arms' length are to be believed. The Spanish braza is 1.67 metres, the Argentinian braza 1.73 and the Portuguese braa (10 palmos) extends to 2.20 metres. Are Portuguese arms really four feet long?

For a wedding feast, it pays to be careful with the condiments. A dash of olive oil is, in the United States, officially defined as six drops, while salt can be tested with a salinometer and pepper checked against the Scoville scale which rates gastronomic to physical temperature (an ordinary chilli rates a mere five hundred degrees, but a hot one soars to eight thousand). The misery of the morning after depends on how strong the drinks, in degrees Sikes, were the night before. Its agony can be assessed in terms of dols (scored from one to ten by comparing the pain experienced with the sensation induced by a bright lamp burning the skin).

Such conspicuous consumption, if it continues – the Laffer curve could be useful here – involves the danger of becoming completely circular (a plane, closed figure with an eccentricity of zero). To avoid such a fate one must watch out for too many of the units of heat in the centimetre-gram-second system, equivalent to a thousand times the heat needed to raise the temperature of one gram of water by one degree centigrade at atmospheric pressure (the calorie).

Those who overindulge at such events face a real risk of a heart attack. It may be treated with the drug digitalis, the amount to be measured by a posologist, or expert in drug dosage, in frog units – the number of live frogs rendered comatose by the appropriate prescription. If he is unscrupulous he might turn to the (as yet experimental and unapproved) heart drugs extracted from poison frogs themselves. A scruple or so (one twenty-fourth of an ounce) should do the job.

The Underground City

The death of a friend in his forties brings thoughts best kept beneath the surface. I stood at the graveside of the traveller and writer Richard Trench as he was buried in Kensington Cemetery, after a life cut short by a heart attack.

Richard's life was one of boundless curiosity satisfied in the best possible way, by finding out for himself. To do so meant a solitary journey across the Sahara, six months as the solitary Unionist in 'Free Derry' and a series of subterranean expeditions reported in his best-seller *London Under London*. Why is the Brigade of Guards under Bloomsbury? Who sang the national anthem to Queen Victoria from the sewers beneath Buckingham Palace? And what became of the plan to pipe their contents to the suburbs, for sale in corner shops for garden use; or of Thelma Ursula Beatrice Eleanor, the first baby born on a Tube train? All this and more is in his account of a vast and curious three-dimensional city unknown to those who live on its surface.

Richard Trench's underground London was in the main a Victorian metropolis. It was built when life was lived in the atmosphere – a layer no thicker than the varnish on a school globe – and when even the deep oceans were thought to be dead. Life in the abyss began with the *Challenger* expedition of 1872, which revealed a hidden world of new creatures beneath the sea. In the two decades since Richard's book life under London – under any city – has undergone a sea change. A strange and rich new universe stretches far beneath our feet, into the Earth itself.

Bacteria are found more than two miles down and, as

some can survive at 130 degrees, there seems no reason why they should not exist another mile below that. The creatures discovered by oilmen ten thousand feet under Virginia are in what was once a stream, buried for two hundred million years under layers of sediment. Its inhabitants have been isolated from the world since the days of the dinosaurs. They are the tip of a subterranean iceberg. Dead bacteria gush from ocean vents, proof that even the earth beneath the deepest seas buzzes with life.

Buzz is not the right word. The Underworld – like much of the sea, but hot where the ocean is cold – is an austere place, where food is short. Some of its denizens use not oxygen but minerals such as iron to burn the carbon they need for food. Bacterial density is a million times less than in the topmost layer of soil; and their cells divide not every few minutes but once in a hundred years. Even so, the new-found land inside the Earth holds one part in a thousand of the entire mass of life, which is impressive for a place unexplored until the 1990s.

Remote though it seems, the buried world impinges on our own. Change and decay can, with the help of bacteria, take place in the deepest mines – which should worry those who hope to use them to store nuclear waste. The bugs might also be persuaded to coax reluctant minerals from the rocks as they digest them. In return, they could be fed with industrial effluent that would be turned to stone.

The difference between men and mice – or between men and oaks – is, relative to the gulf that separates both of them from the creatures beneath our feet, tiny. The DNA of deep-sea bacteria is so different from that of familiar plants and animals as to change the shape of the whole tree of life. It reduced what once seemed its giant branches (on one of which men, mice and oaks reside) to unimportant twigs. Bacteria from far underground, although they have as yet scarcely been studied, already look so distinct that the evolutionary gulf between mankind and the familiar animals and plants that surround us will, in comparison, shrink still further.

The first sign that hominids had become human – and had released themselves from a biological heritage shared with apes or bacteria – came not from genes but from fossils. The Neanderthals began to bury their dead. They marked the graves with ibex horns, the remains, one can surmise, of a funeral feast from which the smoke, as it rose, took the departed to a world elsewhere. Last week we had white balloons with, at the graveside, 'Full fathom five' from *The Tempest*. Richard Trench's soul was borne metaphorically but conventionally aloft by helium as his body – and perhaps his own unique spirit – entered the London under London where, in the end, we all belong.

Post-Modern Physics

The University of California at Los Angeles shows what mass education really means. Bright sunshine, superb facilities, clean students – but still the faculty complain. Too many students (sometimes a thousand a lecture) and too few staff. No room in the library and Students Aren't What They Used To Be (when were they?). Worst of all, some almost never open a book, as they have never been trained to do so.

The campus bookstore shows the size of the problem. It has lots of tedious texts on abstruse subjects but whole shelves of non-books (or anti-books) as well. Nowadays the message is the medium, and the medium is, or so it seems, anything but lines of print on a page.

Biochemists have a songbook, with chemical pathways set to music to make them easier to learn. Those baffled by relativity can join a rocket that hurtles through space on a compact disc and play tennis as it speeds up or slows down. And there is a cartoon guide to every subject under the California sun.

But still, all this is somehow too conventional. More adventurous experiments are needed. One physicist has come up with a radical proposal with a text to match. Students may not read, but they do watch television, much of which consists of recycled science-fiction films. Why not use the vehicle they understand best to teach them what they need? The plan is to use movies to introduce physics. Lie back, relax and some science may seep in with the fiction. Those dreadful old films hide lots of helpful little messages.

In *The Day the Earth Caught Fire* two atomic explosions push the Earth towards the Sun. But how does this fit Newton's Third Law – that action and reaction are equal and opposite? Simple: it doesn't. The Earth weighs around six thousand billion billion tons. The bomb would have to blast a good part of this into space to move our planet in the desired direction. Let's say that it hurls a hundred million tons of debris into the void. The explosion demands a bomb far larger than any ever exploded (and one that would kill every Earthling). Some simple sums show that even this inconceivable blast would shift the Earth just a quarter of an inch out of orbit.

The film is, alas, beyond rescue by science; it is pure fiction. *Aliens*, too, fails first-year physics. The crew saunter about their home as it travels through space with no apparent gravity – but they should, of course, be floating. In contrast, *2001* passes with honours. The giant wheel of the spaceship rotates, which generates enough centripetal force to give those aboard a sense of 'down' and 'up'.

When it comes to relativity, most people are confused. In *Star Trek: The Voyage Home* the ship spins round the Sun and picks up so much speed that it plunges back into history. But why should that cause time to go backwards? Einstein said that nothing travels at more than the speed of light, not that the clocks will run the other way if you go fast enough. *Superman* is even more baffled by the subject. He flies round the globe anticlockwise to swoop into the past and resave Lois Lane. There is a great fault in that. Time is not like a car, for it has no reverse gear.

Physics also puts paid to the Giant Ant Scenario, a useful plot sometimes called in when imagination fails. Ants are ant-sized for good reason. As they have no lungs, oxygen has to diffuse into their tissues. The time it takes depends on the square of the distance travelled. An inch-long ant hence has ten thousand times more oxygen available in its tissues than would one the size of Arnold Schwarzenegger. Such a frightful monster would have to wheeze for breath for some hours before it had the energy to bite the heads off teenagers.

A novel chance to avoid conscious learning altogether has emerged from the new field of sonoluminescence. Pass a loud enough noise through a liquid, and bubbles form. As they collapse they give off bright and regular flashes of light. The information in the sound has been translated into a different medium.

That gives a chance for subliminal study. Encode the data into rap music, play it loud enough and the agitation in the retina should pass the information straight in. It is not a noisy party: everyone is in fact studying for the exam.

The Persistence of Memory

'All is vanity, all is delusion': Tolstoy, on battlefields rather than the BBC, but the films at which its Natural History Unit excels are filled with deceit. David Attenborough does not use stuffed tigers, of course, but all films, on all subjects, are based on trickery, of many kinds.

Some can be forgiven. Telephoto lenses mean that to record sound for a distant event can be difficult; a tiger's bone-crunching noises are hence made in the studio by the BBC's resident bone-cruncher (a busy man nowadays). Some is less easy to excuse. Disney wildlife movies – the high point of the genre forty years ago – relied on simple deceit, with lemmings bulldozed off cliffs and squirrels hurled upwards until a hawk finally caught one.

Most of the dishonesty, though, is inevitable. In a film projected at twenty-four frames per second a stream of single images blurs into the appearance of motion, but why? The standard explanation cites 'persistence of vision': the notion that the next frame comes into view before the impression of the previous one has faded away. That effect does exist. It explains why fireworks are fun, for without it they would be short lines of light rather than fountains of flames. However, it applies mainly in the dark, when the eye adds up photons over time to compensate for the shortage of information available (which is why night-vision shots of hyenas, which work in the same way, smear the image reflected from the animal's eyes). Man has in fact, compared with most mammals, a rather feeble ability to detect flicker and can be fooled at a low rate of frame renewal.

Tigers (or at least their cousins, domestic cats) are much more percipient and would turn away in disdain from a film of a fleeing gazelle.

The truth about film is more subtle. As an object passes across the eye's field of view, it stimulates a receptor in one place in the retina. As the image moves on, it does the same to a second receptor some distance away in the direction of motion. The first signal is delayed by the nervous system as the image moves between the sensors. The messages, one from each sensitive cell, arrive at just the same time in the part of the brain that detects motion, and this interprets their joint arrival as movement at a particular speed. The faster the object moves, the further apart are the sensors from which simultaneous messages arrive, and the more rapidly the brain assumes the object must be travelling.

Ingenious though the mechanism may be, it cannot distinguish between a spot that moves smoothly from the first receptor to the second and an identical spot that is switched off on the journey, to return at the crucial moment. That is the real reason why a flickering series of still images on film gives the semblance of motion. Persistence of vision is in no way involved.

Even so, the brain does persist in error after the film is over. As the credits roll downwards and then stop, the blank screen can sometimes appear to drift upwards. That is because our downwards-motion detectors get tired and fire off impulses at a slower rate when movement has gone on at the same speed and on the same track for a while (which is why a tiger creeps up on its prey rather than pirouetting towards it). The brain responds by assuming a drift in the opposite direction.

Film directors are well aware that they peddle illusions, and try hard to get it right. Dozens of websites of astounding nerdishness list their failures: 'Just before Gollum steals the Ring, Frodo and Sam are sleeping side by side, but in the next shot are at right angles.' The film buffs should not bother, for our visual memory is feeble unless we concentrate hard on a particular object. One droll experiment has shoppers ordering goods from a man behind the counter, who stoops to get them, to be replaced by a quite different man (or even woman) hidden below. Most customers do not notice, which means that a televised tiger could change its stripes between shots with more or less complete impunity.

Film is giving way to electronic tape, which pours out masses of information and does not depend on an image repeated at a speed just below the eye's ability to detect flicker. As a result, as programmes that take the viewer on a walk with a dinosaur or a woolly mammoth show, our ability to fake a scene now has almost no bounds. That may be good news for lemmings and squirrels, but it will be a sad day for natural history when the digitised tigers take over.

Alas Poor Yorick, I Knew
Him Slightly

Stamps and coins are the icons of a nation's history. They are often used to puff its great achievements. One recent Royal Mail effort marked – what else? – the fiftieth birthday of the double helix of DNA, in 2003. A series of cartoons featured a suspiciously chubby and contented scientist wrapped in a model of the double helix. It also salutes another great British strength: a free hand for artistic licence when it comes to science.

The latest production of the Double Helix school has an error; trivial, perhaps, but a hint of a more general problem. It concerns the shape of the molecule, portrayed as a twisted electrical flex with each wire a mirror image of its partner. The truth is less simple and more interesting; for the helix is asymmetrical (as shown on the official Birmingham first-day cover – which will confuse those with the right frank but the wrong stamp). The molecule of life is not a simple spiral, but when viewed from the side has a major and a minor groove. Its shape comes from the way in which the DNA bases are attached to their sugar backbone and affects how proteins bind to genes and control them – complicated, maybe; but it would be good to get it right.

Imitation, as the old joke has it, is the sincerest form of philately and the postal celebrants follow a noble tradition of getting things wrong. The previous DNA stamp, a couple of years earlier, was more elegant but missed out the entire internal scaffold that holds the molecule together – but the image was, said the authorities, Art, which made it OK.

The dubious helices each have a drastic fault and fit into

a sadly consistent pattern. Take the ten-pound note, which bears an image of a bearded Darwin and of the *Beagle* under full sail. It also has a bird which the Bank's own description claims to be a 'humming bird based on the type characteristically found in the region of the Galápagos islands'.

If that were so, the tenner would have a blank space, for although the islands have a remarkable avifauna (Darwin's famous finches most of all) the humming clan is quite absent. Plenty of those birds in South America, but none on the Galápagos; and no mention of such creatures in *The Origin of Species*. The Royal Mint might consider reissuing its note with the image of a Galápagos mocking-bird instead – which would admit the mistake and commemorate a group more important to Darwin's ideas while on his famous voyage than were the finches themselves.

The new two-pound coin also carries a double helix, with a handsome major and minor groove, but the humble day-to-day version of the same coin gets academic egg on its face. On one side it bears an image of the Queen, and on the other a vision of gritty British engineering: a set of twenty toothed wheels meshed in a circle. Impressive, but – alas – also an icon of our problems with technology, for with that set of cogs the system would be frozen into immobility. Only an odd number of gears meshed in such a way will turn at all.

The biggest shocker of all was on the last pound note, the inelegant green billet whose shrunken state acknowledged the long decline of our national currency. Its hero was the Master of the Mint, Sir Isaac Newton himself. His ascetic features peered over a diagram of the solar system as each planet pursues its endless ellipse around our local star.

However, the Sun was in the wrong place. Instead of being at a focus of the ellipse (the point at which rays of light come together when shone into an elliptical mirror) its image was at the centre (where a paper cut-out of the shape would balance when supported on a pencil). Newton's successors of the 1980s had published a map of a universe that would at once spin, rather like the economy of their own day, into chaos.

Newton became a mystic, as have some of his scientific successors; a fact noticed by the Royal Mail. In the presentation pack of the Nobel centenary stamps of 2001, the prize-winning physicist Brian Josephson was quoted as saying that quantum physics might explain 'processes still not understood within conventional science, such as telepathy – an area in which Britain is at the forefront of research'; a statement greeted with dismay by his fellows who deny the very existence of such a thing.

The year 2005 marked the centennial of the birth of C. P. Snow, originator of the notion of two cultures, the arts and the sciences, each opaque to the other. The Post Office did not, alas, come out with a set of stamps decorated with misquotations from Shakespeare to celebrate his life.

A Risk Assessment for Armageddon

Apocalypse is, some say, around the corner. Even for the Elect it will be a hazardous business. In America steps have been taken to minimise the risk and bumper stickers warn: 'On hearing the sound of trumpets, grasp steering wheel firmly.' Our own Health and Safety Executive has not yet added that to its armoury of vexatious little notices. Even so, as the prophecies of the Book of Revelation unfold, its work will be cut out.

The first problem will be the arrival of the Great Beast 666. This aggravating creature is a fiend with seven heads and ten horns that contains within itself the attributes of a lion, a bear and a leopard. God, the greatest genetic engineer of all, has been at work.

As a Genetically Manipulated Organism the Beast falls under a Host of regulations. Before it can be allowed to make war with the saints it must first present itself before ACRE, the Advisory Committee on Releases to the Environment. That governmental body considers any plan to unleash altered creatures (Antichrists or otherwise) on to the British public. Its rules are, as they should be, stringent.

Before a release, blessed or not, the regulations say that there must be a pilot experiment. For Beasts, it might be safer to grant initial permission for just two heads and five horns. Containment is a problem; strict rules for mice and bacteria keep the activists out and the organisms in, but the Great Beast can summon fire from heaven, which may make life difficult for its keepers.

Any proposal to carry out an experiment to end the world demands a risk assessment, an estimate of the chance of Armageddon going wrong. There the Higher Nonsense begins.

ACRE was faced with a scheme to release an engineered virus into cages containing caterpillars on a farm near Oxford. The virus contains a gene for scorpion toxin, which might kill off pest caterpillars ('My father has chastised you with whips but I shall chastise you with scorpions' – 1 Kings 12:10). The risk assessment that calculates the chances of disaster in the fields of Oxfordshire does so in a most eccentric way.

The caterpillar virus has a distant relative that attacks vertebrates, animals with backbones. What is the chance that the released creature could change into a version that might harm us? The virologists work this out with what seems an obvious calculation. Count the differences between the order of DNA bases in the caterpillar virus and that of its relative. Work out the probability of each caterpillar base changing by mutation into its vertebrate equivalent. Then multiply the two figures together.

The logic is that of an engineer in a soap factory. If each of two valves has a one-in-a-thousand chance of failure, then the likelihood of both breaking down at the same time is one in a million. That simple logic gives an estimate of disaster for the virus of, apparently, one in ten to the power of 180. That must be safe – after all, there were only ten to the 150 atoms in the universe when last they were counted.

The Oxford experiment is, all evolutionists know, billions of times more dangerous than that. Their certainty rests on the existence of improbable creatures such as the dugong, the wombat and the drug-resistant bacterium. Each one arose through a process ignored by engineers, genetic or otherwise – natural selection.

Creationists argue (as does the Advisory Committee on Releases to the Environment) that the chance of evolution making an eye is the same as a whirlwind blowing through a junkyard to give a Boeing 747. True enough – if the eye or the aircraft evolved in one step; but it did not. Instead there was natural selection, the slow accretion of unlikely events, each of which improved an eye's ability to see and, as a result, its carrier's capacity to reproduce. The evolution factory makes its complex machines, not with a designer, but through a series of successful mistakes.

Any change in a released virus that might allow it to cope even slightly better with its new situation will spread in the same way. Soon a novel form could appear, improbable though that might seem to those concerned with the dangers of soap manufacture. The chance that the virus will evolve to attack humans is tiny indeed (although it has a better chance of gaining an ability to infect other caterpillars). But one thing is certain: the logic of how genetic engineers assess risk needs another look.

Useful statistical work has already been done. The Great Beast's number, 666, was arrived at in the same scientific way as the infamous risk estimate. If the Emperor Nero's Greek name, Kaisar Neron, is written in Hebrew letters (which are also numerals) one gets *qsr nrwn*, or 666.

And what about Ronald Wilson Reagan – six letters in each name? Unnerved by the coincidence, Reagan changed the street number of his house from 666 to 668. He had an engineer's opinion of biology: 'Evolution is only a theory which is not believed to be as infallible as it was.

Recent discoveries point up great flaws in it.' His most recent successor agrees about Darwinism. As he says, 'The jury is still out.' Engineered viruses may yet prove both of them wrong.

The Madness of Kings

Few politicians escape without accusations of insanity from their electors, and sometimes the electors are right. Many national leaders have taken risks with what the human brain (even one elected to high office) can be expected to accept.

Whole nations have been stretched on the couch – China, say the psycho-politicians, is governed from the left side of the brain (language, rigid thinking) while America is under the control of the far right, in both senses. Many statesmen have been psychoanalysed with the help of that useful Freudian potion known as hindsight. Churchill's 'fantasies of infantile omnipotence' must have been a great help when it came to victory in the Second World War. Stalin, in contrast, had a sense of inferiority that arose from being beaten by his drunken father, which explains why he admired Hitler and had a nervous collapse when the Germans invaded. Mrs Thatcher's problems were with her mother; her triumphs over her enemies, claimed one of her opponents, emerged from her early toilet training.

Such speculations are not much constrained by fact. There are, though, some intriguing figures about the parliamentary mind. Sixteen of the twenty-four British prime ministers from Spencer Perceval to Neville Chamberlain lost a parent when they were young – a ratio thirty times higher than in the general population; and Tony Blair's father had a paralysing stroke when the future prime minister was eleven.

Cognitive dissonance – the deliberate avoidance of information that clashes with one's convictions – is an analytic term with a political flavour. Stalin only once went into the countryside in the years of collectivisation. The Shah of Iran also lived a life dedicated to the avoidance of unwelcome reality.

Psychosis is no more than what one expects from any decent leader. People diagnosed posthumously (and away from the protection of the libel laws) with that condition include Nebuchadnezzar, Cambyses of Persia, Henry VI and Mad King Ludwig of Bavaria (who at least used his folly as an excuse to build a decent folly). The madness of George III is well known; but in addition his Lord Chancellor, Charles Yorke, committed suicide at the time of the American taxation crisis, and Castlereagh, the Minister of War, cut his throat as he swung from elation to despair.

Psychiatric problems are often a symptom of organic disease. Statesmen are as much at risk as anyone else. For neurosyphilis we can turn to Randolph Churchill, Lord Northcliffe, Mussolini and Atatürk. Hitler and Franco specialised in Parkinson's disease, while Stalin went in for pre-senile dementia. Strokes and associated brain damage offer the choice of Woodrow Wilson, Eisenhower, Nehru, Brezhnev and Lenin.

The pathologist who dissected Lenin's brain spoke with a certain admiration of 'an incredibly widespread sclerosis, as if this fossilisation had been the result of some colossal mental labour'. Stalin was impressed. He set up an institute to cut up the intellectual fossil in the hope of finding where its genius lay. Fifty years later, at a dinner party in Los Angeles, I was startled by a venerable Spanish anatomist who showed me a section of that very organ, preserved from his days in Moscow.

If damage to the leader's brain is not enough to spark off an upheaval, drugs may do the same thing. Alcohol led to the fall of Harold Wilson's colleague George Brown, and, so said his enemies, Richard Nixon was unable to carry out governmental business during the Watergate crisis because he drank so much. Churchill, in contrast, claimed that he had taken much more out of alcohol than alcohol had taken out of him. Amphetamines produce symptoms close to those of schizophrenia. Hitler took hundreds; and so did Anthony Eden during the Suez episode, a period when his mood swung 'from feverish tantrums to bouts of almost unnatural calm'.

Other drugs enter the body politic by mistake. Damp bread grows mould. Mould contains ergotamine, a relative of LSD, and LSD leads to bizarre behaviour. The *Marie Celeste* was found deserted; perhaps because the crew ate mouldy biscuits and, pursued by imaginary demons, leapt overboard. That seems convincing enough – but could both the Russian and the French revolutions have been sparked off by the mass ingestion of putrid food, as psycho-historians claim? Each happened, it seems, after long periods of rainy weather. The case looks watertight.

Whatever the psychiatric crises of past leaders, one thing is clear. No one could accuse today's small group of potential prime ministers of being anything other than morbidly sane. Perhaps now is the time to despair.

It's Not Cricket

Stephen Jay Gould once claimed to have found a tie between the history of baseball and that of life. It resulted from a careful analysis of the game's scores over the past century.

High hitters are a thing of the past. The record average score was set as long ago as 1941 and in the decades before then many players did almost as well. Since Pearl Harbor, though, such impressive tallies have disappeared. Most fans blame a decline in the quality of the game.

For some biologists, too, evolution means only improvement. Life is arranged in a pyramid of perfection; at its base live bacteria and single-celled creatures, then sponges, insects, birds and mammals and on its summit the human species, monarch of all it surveys.

Such views reflect a shared delusion because they turn on exceptions rather than the whole story. A billion years ago most organisms were bacteria – and they still are. Far from a general advance, evolution has, in general, been static. The horse – often painted as a triumph of progress from terrier-like creature to Trigger – is in fact a remnant of a group that once had dozens of species. We, too, think ourselves special, but – horse or human – we are but froth on the Age of Bacteria. Baseball is just the same. If, instead of picking out the extremes – the high hitters – one analyses all the scores, a remarkable picture emerges. Like life itself, there has been little *overall* improvement since the sport began. Annual averages over all players are, in general, stable.

There has, though, been a shift in the spread of scores.

Variation among players has long been in decline. Once, some were so much better than their opponents that they made hay of them. A poor pitcher meant that a decent hitter scored off every ball; a poor hitter would strike out. That is no longer true. Today's lack of record-breakers is not because the game has deteriorated but comes from increased competition. Everyone has improved, and as nobody – pitcher, hitter or fielder – can stand out from the crowd the old records remain unchallenged. Nowadays fielders succeed in 99.7 per cent of attempts to catch the ball: there is simply no room to get any better. Evolution in the baseball world has led not to progress but to a struggle for a place at the edge of what is physically possible.

Gould's law applies much more widely. Skyscrapers face the Empire State Problem: progress is fast at first, but slows down in the face of the laws of nature. The Flatiron Building in New York was built in 1902. At three hundred feet, it was the world's tallest structure. A brief seven years later came the Metropolitan Life Tower, twice as high. It took a quarter-century for the global record-holder to gain a quarter more in altitude (the Chrysler Building at 1048 feet, erected in 1930) but we had to wait until 1973 for the ill-fated World Trade Center to gain the next quarter or so.

Taiwan has topped out a record-breaker at 1667 feet, just a little higher than the fallen New York giants. The next winner will be in Dubai at 2313 feet, taller than Taiwan's erection, but 2009 will certainly not be like 1909, with a modern Met Life that rises to twice the height of its rival. Architects have an inexorable enemy called gravity which, like a lethally effective baseball catcher, takes no prisoners.

Armed with sixteen hundred pages of *The Wisden Book of*

Test Cricket I checked whether Gould's law fits our national sport. For each of 548 English players since 1877 I took a lifetime average batting score. Twelve born before the First World War had an average above fifty; a distinction achieved by just three of those born after it. The famous W. G. Grace, with his average of thirty-two, was low in the lists. R. H. Ward (born 1853) did more than twice as well, with an average of sixty-eight (never since surpassed).

In one way the story is the same as in America. The average number of runs per Test was 18.3 for cricketers born in the 1840s, in the 1890s 17.2, in the 1940s 21.5 and in the 1960s 18.5. As in baseball, cricket shows no sign of any overall progress in average score. Great players still walk on to the turf – but to concentrate on those exceptions misses most of the story.

What about an increase in efficiency through competition? The figures on this side of the Atlantic reflect, alas, our national weakness. Baseball has got leaner and meaner over the years, with the spread in scores reduced by nearly half since 1870. Cricket shows no sign that this grim logic is at work. The differences among players today are just as great as they ever were. Far from rivalry and an escalation of excellence, with everyone pushing against the wall of what is possible, our national sport is in an evolutionary dead end.

There is one big difference in the evolution of the two games. Baseball has always been professional, with big bucks for both players and managers. Cricket, in contrast, is (or was) run by gentlemen and can afford to take it easy. What this tells us about the psyche of the two nations is hard to say.

Bringing up the Rear

Thanks to global warming, spring comes earlier each year, but Londoners cannot afford to lift their eyes to the cherry blossom. Instead they must keep them fixed on the ground to navigate the great reefs and promontories of canine mess that impede the city's populace on its way to work.

Much though most pedestrians would like to rub the face of every dog owner into his animal's calling card (and some pets are either the size of a small horse or eat buckets of brown rice) the science of scatology does not, unlike London's pavements, get the coverage it deserves. As the Court Physician in Alan Bennett's play *The Madness of George III* points out: 'I have always found the stool more eloquent than the pulse' – but who now sings of stools?

Many biologists do little else. I once went to a talk on the joys of the African diet as interpreted from the end product and although the slides left an indelible impression it did persuade me to shift to bran rather than bacon for breakfast. But what should be the natural state of our bowels? Plenty of books on the palaeolithic diet suggest that nuts, berries (and perhaps the occasional avocado) are the true foods of the human race and will lead to intestinal ease. However, coprolites – fossil droppings – tell a more ambiguous tale. They unearth the supply side of the ancient digestive equation.

Those deposited by Neanderthals reveal that they ate an almost vegetable-free diet, while the relics of their successors – our own ancestors – hint at a move towards fruit and fibre (whether a shift to the Islington Breakfast drove human evolution is harder to say). Fashions often change. Muscle proteins in the petrified dung of twelfth-century Pueblo Indians show that they had turned to cannibalism and the finest flesh of all, while the flakes of gold leaf in the loos of Pompeii hint that a rich diet would have done for that decadent race even if global indigestion in the form of Vesuvius had not put paid to them first.

Animals, too, speak with a certain eloquence from their backsides. The blue whale excretes a series of clues about its sex, its parasites and its diet in a fine brown cloud, several times a day. In an attempt to reduce Japan's 'scientific' whaling quota (in which animals are killed to obtain samples, before their meat goes on sale in Tokyo) the International Whaling Commission plans to use the DNA found in whale excreta to ask questions about the genetic structure of populations. The Japanese remain unimpressed.

As top predators, with a high-protein diet and an efficient colon, whales do not put much back into their environment in the form of dung. Other creatures, though, repay their debt to nature in a more generous fashion.

Darwin noted as much, with his 'worm-stone' at Down House in Kent. The heavy millstone he placed on the surface has, over a century and more, sunk several inches into the ground as silent witness to the strainings of the earthworms below as they pass undigested material – eighteen tons an acre each year, he calculated – through their guts.

Full marks for effort for the Kentish worm, then, but its tropical kin pumps out twenty times as much. Even those labours pale in comparison with what goes on in fresh (or what passes for fresh) water. Queen Victoria, while crossing London Bridge, asked what all the bits of paper in the Thames might be. 'Notices that prohibit bathing, your

Majesty,' replied a quick-witted courtier. The river is far cleaner than it was (the old joke about not swimming but going through the motions no longer applies) but new work shows that even the clearest mountain stream is full of waste.

Black-fly larvae live in fast rivers and filter out tiny particles of food from the water. Their digestive systems are so inefficient that they absorb just a fiftieth of what they eat. As a result, such creatures pump out lots of bulky faecal particles that sink to the bottom, where they stay until they are eaten by other animals. At the outlet of a certain Scandinavian lake, the billions of wriggling worms return more carbon to the system via their heaving guts than does any other creature; and, without their labours, huge amounts of dissolved nutriment would be washed straight out to sea rather than circulating through the food chain. Insects and fish would starve and the rivers would become more or less sterile.

Even London's *merde* has its uses and in Victorian times the poor scraped the pavements to pick up dog waste for sale to tanners. If whales can be identified from their excreted DNA, dogs no doubt could be tracked down in the same way. Then those who allow them to foul the streets might be dissuaded by a quick DNA fingerprint followed by a heavy fine. Or do I mean footprint?

A Tide in the Earth's Affairs

London Bridge at high tide on a spring day is a cheerful place to be, with the sun sparkling on the water and the memory of Samuel Johnson's famous attack of river rage at this spot: 'Sir, your wife, under pretence of keeping a bawdy house, is a receiver of stolen goods!'

It makes more sense to feel depressed; not in spirits, but in body. When the tide is in, London sinks under the weight. Not much, but each Londoner comes down slightly in the world as a result. They rise again in a few hours; but each day a tiny sag of the Earth's crust beneath its watery burden reminds us that we live in an elastic world.

Since 1992 the satellite Poseidon, named after the god of the sea, has scanned the surface of the ocean. Poseidon is accompanied six miles below and one minute ahead by his planned successor, Jason (of Argonaut fame). The mechanical deity and his heir fly a thousand miles above the surface. They weave an endless web of observations as they cross and recross their own track and map the entire ocean once every ten days. As the satellites race through the skies they send down radar pulses that measure the reflection from the ocean surface. The delay between signal and echo tells the height of the sea to within an inch or so – which is equivalent to the thickness of a penny viewed from a transatlantic jet.

Poseidon's course is affected by gravity and, as a result, by the mass of the land and water over which it passes. As the tide below rises, the satellite is nudged a little closer to the Earth. However, a close comparison of Poseidon's bounce with the height of the ocean below shows a tiny disparity. The satellite seems to respond too much to the rising sea, for it sinks lower than expected from the extra mass of water as the tide comes in. Something else must be shifting it from its path.

Clever mathematics reveals what is to blame. The extra deviation is because the Moon causes tides deep within the Earth itself: not just the weary sag as the waters cover the ground, but a movement in our planet's very rocks. Poseidon, master of the seas and streams, was also god of earthquakes (which in ancient times were thought to be caused by underground rivers). Although the Moon's effect is far too small to spark off such an event, the Greeks at least gave him a sphere of influence on land as well as in the sea. The solar mass also has an effect on the more prosaic oceans here on Earth. When the Moon and Sun line up, their joint pull ensures a greater deformation of the seas – a spring tide – than when they are at right angles. The Earth, too, stirs a little more than usual.

The tides within our own planet are trivial compared with those inside the Sun itself. Our local star is in a constant state of agitation, with more than ten million distinct modes of vibration. Most are not tides in the strict sense as the influence of the planets on their parent's mass is small, but their presence means that the source of our heat, light and very being reverberates like a struck bell.

Such resonance is under investigation by a set of observatories called, inevitably, the Global Oscillation Network Group (GONG). GONG shows, for the first time, that the Sun has a more or less rigid centre, covered by oceans of fiery liquid. Just as in boiling porridge, thousands of close-packed cells on the Sun's surface transport hot material as it simmers up from below. At the equator massive tides wash back and forth but towards the poles they are much reduced. As a result, hot material at the surface flows from the equator, to sink inwards at the Sun's northern and southern ends. Another unexpected tidal flow arises because, for some reason, the fluid surface of its northern hemisphere spins slower than does the equivalent in the south.

On the way home from London Bridge, I make the mistake of hailing a taxi in the Strand. When it comes to the ebb and flow of traffic, Dr Johnson had it right: 'Fleet Street has a very animated appearance; but I think the full tide of human existence is at Charing Cross.' As the Thames drains away, oblivious to the stationary wave of vehicles on its banks, the time has come for an attack of road rage at a cab-driving descendant of Johnson's insolent boatman: 'Sir, your wife . . .'

Why Not Eat Insects?

There are, said Georges Clemenceau, only two useless organs: the prostate gland and the presidency of France. He was wrong about the role of the gland, but that of the President has swollen. In the election of 2002 the nation faced what many citizens saw as the worst of all possible worlds – Le Pen versus Chirac, a choice between two deeply unattractive options. In the end the leftish masses held their noses and turned – as they said – to the crook rather than the fascist.

Many creatures face such disagreeable decisions, not about votes, but about lunch. Many items are nasty, or expensive, or hard to find. Food is governed by two economic truisms: 'Never eat anything bigger than your head' balanced against 'Beggars can't be choosers'. A morsel that is tasty but too hard to handle is not worth the effort, but in tough times it is worth checking out what is available even if it smells a bit off.

Most people are conservative when it comes to voting with their teeth: Vincent Holt's forgotten work of 1885 *Why Not Eat Insects?* was defensive from the start ('In entering upon this work I am fully conscious of the difficulty of battling against a long-existing and deep-rooted public prejudice'). It contains recipes for *larves des guêpes frites au rayon* (wasp grubs fried in their nests), *cerfs volants à la gru gru* (stag beetles on toast) and many more.

And why not? Hunter-gatherers ate dozens of different things, and one Aboriginal group in Queensland was recorded as eating two hundred and forty kinds of plant.

Nowadays the top five food plants for all the countries in the world, each with its national preference for chips, or for fried rice, involve just half that number of species.

There are still plenty of national contrasts in diet. Voltaire said of the English puddings (and plays) of his time that 'nobody has any taste for them but themselves' – but today a greater variety of cuisine can be found in any English city than in the whole of France. *De gustibus* – it seems – *non disputandum*, but in reality there exists a whole science of accounting for taste.

Individual differences in preference are everywhere. Experiments with stomach pumps show that some wild trout go for worms, while other trout in the same stream have a taste for gnats. In the same way individual birds presented with a choice of green baits or brown each concentrate on one colour or the other. Some gull chicks will starve almost to death before they can be persuaded to switch their preferred food colour – and the preference lasts for life.

It all turns on efficiency. For a hungry gull facing the same limited menu every day it pays to specialise rather than to search for novelty. Now and again it may be worth switching in the hope of finding a new product on the market, but most of the time it pays to stick to what you know, for that is almost always available. Pet cats can be choosy – and they can afford to be, for however strict their preference their owners will provide it. The semi-wild cats who stalk the streets have a harder time and are far more ready to try a new foodstuff than are their effete domestic cousins.

Many birds face a constant shift in what food is available from day to day, or from season to season. They show the power of novelty. Blue tits can be trained to tear open envelopes that contain delicious pieces of almond. Then a trick: a few of the almonds are dosed with bitter quinine. The tasty packets are marked with small crosses, the rare noxious ones with large and conspicuous squares.

At first the birds' behaviour seems as perverse as that of the French electorate. Horrible as their contents are, the rare but eye-catching envelopes attract *more* attention than average; and the scarcer they are the more interest they generate. Not until the frequency of the poisoned ballots rises to a fifth of the total do the avian voters begin to avoid them. Blue tits, it seems, have short political memories, for the desire to sample a new and flashy item in the hope that it might pay off soon causes them to forget quite how unpleasant their experiences were on an earlier try. In their uncertain economic climate, it pays to gamble.

The same is true for our own shifting financial state. In the hungry days of the Industrial Revolution the *Livestock Journal and Fancier's Gazette* published an article entitled 'Eating Cats in West Bromwich'. *Why Not Eat Insects?*, printed in the same era, also makes the fiscal case for a shift in menu: '*How the poor live!* Badly, I know; but they neglect wholesome foods, from a foolish prejudice which it should be the task of their betters, by their example, to overcome.' In these days of junk food the poor's poverty of imagination is poisoning them once again. Their betters must do better: who would like a pudding of *phalènes à l'hottentot* (moths in butter) or *crème de groseilles aux némates* (gooseberry cream with sawflies)? Eat your heart out, Voltaire: for we can wash the dish down with the Antipodean wine whose bouquet is best described as cat pee on a gooseberry bush.

The Dog It Was That Died

The Nakhla meteorite is black, heavy and – like many such icons – oddly unimpressive. It has another distinction, as it is still stained with the blood of the Egyptian dog killed when it landed on the unfortunate animal in 1911. The hound had, although it had no time to realise it, been struck by a piece of Mars. Most of the time that fragment of the Red Planet resides in a museum in Adelaide, but now and again the object goes on tour.

Nakhla is one of just thirteen Martian meteorites ever discovered. Unlike all other pieces of galactic rock, the chemistry of each shows that at some time in its past it passed through a volcano, and did not – like most meteorites – land straight from space. The gases trapped inside prove just where they underwent their fiery overhaul. The Nakhla meteorite was struck off Mars by a giant collision and blasted into orbit, to land, much later, on an unlucky African canine.

If stones can move between planets, why not germs? Perhaps, some suggest, life originated elsewhere in the solar system, or even farther away, and was carried on such objects as they flew from their native land to Earth.

The oddest thing about life is how soon it began. A mere half-billion years after the Earth spun into existence, it sent out its first hesitant shoots – and not for another two billion years did it make the first proper cell. The standard view is that the first life on Earth arose locally, in a warm tropical pool full of chemicals. If that is true, the early inhabitants of our planet must have had a disturbed childhood, because at the crucial moment, three and a half billion years ago, the Earth was under bombardment by meteorites big enough to vaporise the ocean and flooded with enough ultraviolet to sterilise an elephant. The genes show that the members of the primeval SLIME (the Subsurface Lithoautotrophic Microbial Ecosystem; one-celled creatures that live within the earth) sit near the base of the universal evolutionary tree, so the idea that life started beneath our feet makes sense.

Conditions below the surface are much the same on Earth and on Mars, and both contained all the ingredients for a primeval minestrone. There was no shortage of water on the Red Planet; indeed, its surface was being shaped by glaciers until just a few million years ago and in one location are the remains of what may have been a great lake the size of the North Sea. One of the Mars Rovers has found an iron-rich salt, rather like that found on Earth in places where water soaks through certain rocks. On Mars, as on Earth, the interior is warmer than the surface and there may still be liquid water just a kilometre or so down. If such conditions could foster life here, why not on Mars?

There might, perhaps, have been single-celled Martians hidden deep beneath its surface. But how could they take a trip between planets? Surely, they would fry, freeze or die of old age on the way. However, the idea looks rather less bizarre than it did.

In the 1960s the *Surveyor* mission sent cameras to the Moon. A few years later the instruments were picked up by the *Apollo* voyagers and returned to Earth. Pieces of their foam containers were put into a nutrient broth, and – to the dismay of NASA – turned out to carry bacteria deposited from the noses of those who made them. They had stayed healthy on our satellite for two and a half years.

The preserved sneezes of the NASA technicians are a reminder that life is tougher than anyone thought. Earthbound experiments agree. The Chinese have found that the seeds of plants such as peanuts or rice need only be dried hard with silica gel or quicklime to survive for ten years (which means that the expensive paraphernalia of a seed bank, with its freezers and liquid nitrogen, is not needed). More remarkable, mouse sperm dried to a powder can fertilise eggs when injected into them, with a success rate of nine in ten. There seems no reason why the sperm should not last for years, or even centuries.

The existence of distant cousins of *Homo sapiens* on Mars is no more than speculation; but it is a pleasant thought that President Clinton, Bush or one of their successors could, with a small gesture, father the future child who will grow up to make the first journey to that planet.

Art and Illusion in Alaska

The sound of the lute is everywhere in the land, at least in places where scented soap is sold. Even in the remoteness of Alaska it can be hard to escape it. Alaskan life is a soap opera with a Wagnerian set: black mountains in the distance, bald eagles overhead – and suburban America looming large in the foreground. The souvenir stores of Ketchekan, like those of Stratford-upon-Avon, echo to the plangent tones of five-hundred-year-old Muzak.

Most of the shops sell Indian artefacts. The paintings are attractive but odd. The animals – bears, wolves and the like – look as if they have been run over by a bus (the bookshop sells a *Field Guide to Road Kill*, with outlines of beasts squashed on the highway arranged for easy identification). All four legs are visible, as are both eyes, the nose and the tail. It seems an odd way to paint an animal but to the artists it makes perfect sense. All the relevant bits are shown, which means that it is easy to identify each species in the painting.

In fact the Indian way of showing a three-dimensional beast in two dimensions is no more artificial than are the familiar norms used by Western art. Until not long ago no native Alaskan saw many straight lines. Their artistic conventions hence do not depend on the laws of perspective which seem so natural to us; indeed, so unused were they to the idea that they once found it hard to make sense even of a photograph.

The whole of representational art, of any school, depends on the fact that we are fools tricked by our own brains. Even worse, we must learn to be fooled, and to be fooled in the same way each time. We use faith to persuade ourselves of lies; and – as usual where faith is important – different lies take hold in different places.

How did the brain gain the belief that a flat and imagined world on a piece of canvas can represent reality? The familiar laws of Western painting began just five hundred years ago. The Italian Renaissance polymath Brunelleschi, builder of the great dome of Florence Cathedral, was the first to put life into perspective. He realised that if one drew distant objects smaller than those in the foreground they would appear to be farther away. As long as vertical and horizontal lines are retained in the image, a spectator can be persuaded that a line shown at an angle is in fact receding into the distance. The brain is happy to play games with itself, but first it has to be taught the rules; and – as a baby's picture book makes clear – young children find it hard to use the cues of size and distance that seem so obvious to adults.

Most of us are so used to the conventional laws of perspective that they seem flawless. In fact, and just as in the Road Kill school, they suffer from real limitations. To paint an object without distortion it must be looked at from a distance of at least ten times its own diameter – which means about fifteen feet if the object is a person. Most of the time we are closer to the individual we are speaking to; and every close-up portrait is distorted as a result – a weakness in the laws of Western art that Westerners have learned to live with.

Those damned Magic Eye books that infest today's bookshops use an entirely new artistic code, based on thousands of brightly coloured spots repeated to give a false image of a three-dimensional world. I have tried to learn that artistic language and once fleetingly saw a giraffe but, in general, I look upon such objects as a nineteenth-century Alaskan may have looked at a photograph; or, to quote Charles Darwin, 'as a savage looks at a ship, as at something wholly beyond his comprehension'.

Sometimes differences between artistic codes lead to absurd mistakes. There was once a claim that ancient Egyptians had domesticated the dwarf mammoths that flourished in Siberia (and perhaps even in Alaska) five thousand years ago. The evidence was an image, found in a tomb, of a small but perfectly formed elephant led on a chain by an Egyptian soldier. Surely this was proof that the Egyptians traded with the far north.

A brief but lethal response put matters right. In Pharaonic art the size of an object shows not how big it is, but how important. Slaves were painted smaller than princes – and no one cared about mere elephants. So much, then, for the myth of the wandering mammoth. Maybe a picture of one is hidden in this Magic Eye picture – or is it a squashed elk, or a poached egg?

Why Memes Stay at Home

Scientists are often instructed to tell the public what they are up to; but in general they do not. Instead they talk in language that even colleagues outside their own field cannot understand. Richard Dawkins' ingenious idea of memes, units of knowledge that pass freely from place to place and multiply like living creatures to bring information to those exposed to them, has been much over-interpreted; but if such things exist it seems clear that science memes are – on grounds of simple obscurity – distinct from all others.

That is odd, because the whole point of science is to communicate. No breakthrough means anything until somebody has been told about it. In science, publications count (and are counted). To write a scientific paper used to be a journey without maps. The traveller wandered off to the library to track down the references that link his own discovery to the world outside and, with luck, draw the attention of colleagues to his work.

Nowadays we have the internet and those great Columbuses of knowledge, the search engines. They provide a veneer of culture at the touch of a button. Type in any word and the most unexpected links appear. This, surely, should tie science to the sum of human experience in a new cartography of universal information.

What they show most of all is that the web is not worldwide at all, but consists of a series of overlapping kingdoms that rarely speak to one another. The electronic explorers – Yahoo, Google and the rest – all look in different places. As a result, even the best web search trawls but a fraction of the database. To navigate some parts of the web is like using a Tube map to travel the streets of London; the fit between chart and reality is vague at best.

Lots of mathematics has gone into understanding the structure of the web. The programs that hunt down who chats to whom show that, just like scientists, people stick to their own community and do not often venture outside. Experts talk in learned terms about connectedness and network nodes and make charts of packets of data as they zoom across the globe; but, as it happens, the meme itself tells the story without maths.

Memesiologists have a universal interest in using the internet to test the penetration of their idea. Dawkins confesses to having typed his word into Google. He came up, he says, with half a million responses. The meme meme has, it seems, spread throughout the universe of understanding. A closer look shows just how that cosmos is subdivided and how the meme – like most of us – prefers to stay at home.

Scientists use special search engines that comb through the technical literature, rather than the general library searched by Google. To enter 'meme' into one of those mighty machines is to cast a certain doubt on just how well the term has broken loose from its own community.

Over the past decade the giant electronic Science Citation Index (which lists how often particular papers have been referred to by other authors) has come up with not half a million but just seventy-one hits on that word. They inhabit several distinct republics. A third of the papers refer to a polyurethane breast implant called (for reasons unexplained) a 'Meme' and a dozen to a variety of publications with '*plus ça change plus c'est la même chose*' in the text (scientists like hackneyed phrases although one has '*la même guacamole*' instead). Some refer to a technical term in metallurgy, one to a district of Cameroon and its remarkable damselflies, and one to a gene in cotton. The meme idea itself merits around forty mentions, many from students of birdsong; but some from computer buffs who use the notion to look for themes in programs. The meme meme is, it seems, not much at home in the world of science.

Many scientists have the guilty habit of looking themselves up in the Citation Index to test how often their work has been quoted and how famous they might be. If I do that I receive, for a lifetime's work on snails and fruit-flies, just 654 responses; not a cause for self-congratulation, but after all, who cares about such things?

In contrast, to type 'Steve Jones' into the memosphere explored by Google, gets – praise be! – 31,799 reactions. Fame, it seems, at last; but, alas, that two-word meme is a universal title. There is the golfer, the footballer and the member of the Sex Pistols. Then a beefy bodybuilder with a purple posing pouch (my students like that one); and, at last, the famous author himself – of that best-seller *For Men Only: Winning at the Dating Game*.

Catching Eyes and Throwing Voices

Educating Archie must be the oddest radio programme ever made (for the young among us, its format turned on a dialogue between a ventriloquist and his dummy). Given that neither player could be seen, how could one tell whether the non-dummy was doing his job? Even so, by some strange mental process, millions of 1950s listeners persuaded themselves that Peter Brough really did throw his voice and that Archie Andrews moved his wooden jaw up and down in response (he was believed, by true acolytes, sometimes to waggle his eyebrows as well).

Ventriloquism relies on the fact that we hear with both our ears and our eyes, for it is easier to understand what someone is saying if their lips, or their faces, can be read at the same time. A dummy makes such a show of its jaw movements and eyebrow waggles that we are fooled into believing that he, and not his rigidly static master, is the source of speech.

Speech turns, in other words, on a combination of signals. We can manage with just one (which is how radio works), but do better with several. Both the listener and the speaker sense that. The modern student pastime of tape-recording lectures (it beats all that tedious note-taking) means that lecturers themselves, who are subconsciously aware that those at the back are too far away to read lips, are horrified to hear not their usual modulated tones but those of a thespian hack in full flow.

Signals also combine in the animal world. Males of the dart-poison frog defend their territories with loud croaks and distended vocal sacs if a challenger gets too close. Play back the croak and the jealous frog gets irritated, but not nearly as irritated as by a real rival. An experiment in the jungles of Central America with an 'electromechanical model frog' – a high-tech version of Mr Andrews – shows that to cause serious annoyance to a frog takes the amphibian equivalent of waggling the eyebrows: croak plus bulge does much better than croak, or bulge, alone.

The poisonous little beasts would enjoy *Educating Archie*, for frogs can be fooled with the same kind of trick. Fiddle with the controls and the mechanical Kermit can be made to blow up his throat and then make a croak a fraction of a second later. The defender calms down at once, for to have an effect the two cues must arrive at just the same time. Get that talented but artificial beast to throw his croak by shifting the loudspeaker a few inches to one side, though, and the angry amphibian at once attacks the model, which it assumes to have made the sound.

People, too, are open to such a deception. In a darkened room with a half-circle of speakers, play a beep and flash a light above the source of the sound. The subject points at where he thinks the noise comes from. Shift the light away from the correct speaker and he tends to keep pointing at the light. Repeat that a few hundred times and then, the crucial test, do the experiment in darkness. Now he has been ventriloquised and points at a speaker displaced several degrees from the real one. The delusion persists for half an hour and works even when the volunteer is told to ignore the light altogether.

Science has come up with some helpful hints for those who aspire to the art of belly-speaking. The eyes prevail over the ears when it comes to establishing the point of origin of a sound, which is why Archie and his fellows have big heads, wide mouths, and improbable eyebrows. It is hard to persuade anyone that a sound has been displaced by more than fifteen degrees from the true source. As a result, it makes no sense to have the dummy across the room: the traditional hand up one's partner's backside is still best.

Just as for frogs, timing is important, and a shift of a mere tenth of a second between sound and the model's lip movement spoils the illusion. Our brains can correct a gap between the audible and the visible if it is not too great. If sound arrives a fiftieth of a second before the mechanical lips move, nobody notices. If the lips move first, we can be fooled even when the delay is twice as great. What is more, the trick works best when real and mechanical partner each use sounds of the same frequency (which is why male ventriloquists do not have female dummies).

Even with all this help, most of us can never hope to make it as the Peter Brough of the twenty-first century. However, we know some tricks that he did not: play the sound 'ba' while watching someone silently move his lips to the sound 'ga' and, bizarrely, one hears the intermediate, 'da'. Dottle of deer, anyone?

The Norfolk of the North

Very flat, Finland; a nation that – like the Norfolk Broads – is a dilute solution of land in water. It is good for boats, birds and sauna baths but is not the best place for spectacular views.

Finland, though, has a scenic but invisible past. It was once a lofty mountain range, thrust up during a crash between two great continental plates as they drifted into each other and – like a slow-motion road accident – hurled a mass of crumpled rock into the sky.

In the great scheme of Earth history, mountains are fleeting visitors. The continents move fast – India is penetrating China at twice the rate that fingernails grow – and as they meet may throw up huge peaks at great speed (the Himalayas, for example, are younger than the last dinosaur, at around forty million years old). In geological terms the new additions to the landscape wear away almost as soon as they arrive. Their temporary state makes great mountains rare, albeit popular. On the average the world is more like Finland than Switzerland, with the mean height of the land just a few hundred feet above sea level.

The surface of the Earth has broken and re-formed many times. Some of the evidence can still be seen (the long line of the Ural Mountains marks an ancient crash of plates) but most of it is underground. Finland's bland surface itself conceals a rugged history, hidden far below. The first hint of the existence of a subterranean landscape came from the pioneers – Sir George Everest included – who mapped the world. Away from the plains, their survey grids

and measures of height did not match up – because, it tran-
spired, the plumb lines used to establish the vertical at each
point were attracted away from the upright by the mass of
nearby peaks. A closer look at patterns of gravity showed
that the mountains themselves rest on deep foundations.

When two plates collide as they skim across the liquid
Earth, the mass of material pushed aside thrusts downwards
as well as upwards. Mountains are, as a result, like icebergs.
They float in a sea of rock, with most of their mass beneath
the surface. As the rain wears them away, the roots float
higher, to renew the range from below. In time (and just as
in an iceberg) the peaks are washed into the sea, and the
foundations follow on.

In a young range – the Himalayas, for example – the
roots are deep indeed. They may be ten times thicker than
the summits above (and Nepal's internal Everest is forty
miles below the surface). In older places such as the
Appalachians they are shallow, further proof that the foundations are indeed buoyant and move upwards as the land
above is eroded away.

A map of the buried Earth based on a global gravity
survey comes up with a surprise. Some places which have no
mountains or even hills – Finland is a prime example – still
have their ancient foundations beneath the surface. Just as in
a mouth left to rot without a toothbrush, the mountain roots
remain even when nothing is left on the surface. The great
mass of rock beneath Finland is the relic of a huge range. It
disappeared long before the Himalayas even began. The
holidaymaker may be bored by the view across Helsinki, but
he is standing on the summit of a vanished Everest.

The Finnish Alps soared into the sky and poured into the
seas a thousand million years ago. Their relics were saved

because they became denser with age and, as a result, did not float upwards but sank deeper into the mantle. Finland's basement has cooled down since the heady and heated days when the ancient plates ground into one another and has turned into a new and massive kind of rock that has no wish to heave itself into view.

Lost Himalayas also once sprang from the broad plains of eastern Canada, the open veld of South Africa and the flat and arid pastures of New South Wales. Geologists are much interested in these vanished peaks, for mountains – even those long gone – are excellent places to find gold and other minerals. Finland itself has plenty of zinc, silver and chromium. Like the archaeologists who pillage the buried cities of Greece and Rome, those who study the archaeology of once-great peaks hope that treasure has been left in the ruins even when the edifice itself has collapsed into its own basement. The atlas of the vanished Earth tells them where to dig.

To the traveller in a flat country, the idea of a forgotten mountain beneath one's feet might add interest to an otherwise tedious outlook (although nothing can rescue Norfolk, which is just mud).

Real Science or Radioactive Pasta?

The European Commission in Brussels, with its mounds of food, mountains of paper and towers of Babel, holds occasional seminars on Science Communication in the Media (or, Why Do People Hate Us So Much?) with a pan-European mix of scientists and journalists in desperate search of common ground.

Not everything about the public perception of science is gloom. In the north the subject gets the coverage it needs, and perhaps even more. Germany has science magazines aimed at teenagers, while parts of Eastern Europe also take it seriously (as mathematicians say, 'If I have seen further it is by standing on the shoulders of Hungarians').

The south is the real problem as Greece, Italy and Spain have almost no real science coverage in newspapers or on television. In the lands where the lemon trees bloom, sensationalism is the rule. There has been a great failure of communication; as an Italian delegate put it at one such meeting, scientists don't realise the importance of lunch. Long ago German botanists irradiated the hard wheat used to make pasta to cause mutations and bred from the best plants, with modest success. The story mutated in the Italian press into a scare about radioactive spaghetti as part of a Hunnish plot to drive out the national dish in favour of bratwurst.

There is a deep division between the two sides in their understanding of the other's trade. Reporters complain about the scientists' failure to understand the pressure of time and a newspaper's need for instant response. The press was outraged by demands in a draft British 'Code of Practice' that – as in science itself – independent experts should be allowed to read copy before publication, or should even write headlines. Politicians or murderers would, journalists point out, much appreciate such a privilege. The workers at the laboratory bench, in return, accuse Fleet Street of acting like the Spanish Inquisition; the less it finds, the more it thinks that something has been hidden, and the tighter go the thumbscrews.

As a result, paranoia rules. Journalists may be suspicious, but many scientists see newspapers as monsters to be fed at regular intervals to calm them down. The chemical industry has, rumour has it, a few sacrificial lambs lined up; minor lines of research it will abandon on ethical grounds if the Beast of Fleet Street growls too loud. Perhaps biology, too, should prepare some GM Dolly ready to have her throat cut in public to assuage the tabloid thirst for blood.

The two trades have, over the past few decades, travelled in opposite directions. Medicine was once based on a series of stories about single cases, with no attempt to look at all the evidence. It took the development of statistics, the most dismal of sciences, to escape from their impasse. Nowadays doctors need large samples, objectively tested, and accept that all statements come with a degree of uncertainty.

The life sciences have moved from the particular to the general, but journalism has done the reverse. Where once the press spoke in expansive terms about the wide issues of the day, on medicine as on other matters, now it focuses on individual tragedies. Their view is natural, but it is not science. The family of a CJD victim, or of an autistic child, has a story that deserves to be told, but they are not experts; and those who are resent the way in which a mass of technical evidence is given just the same weight as the anecdotes, painful as they may be, of those who have untested theories of their own.

What is to be done? Perhaps the Commission could train scientists to go to the media rather than vice versa, or invite journalists into labs (risky, given the usual mix of sealing wax, string and high explosive); or set up a database of communicators, or an office for science information, or science firefighters ready round the clock to refute the scares. Most have been tried, with mixed success, and perhaps should be tried again.

Science could turn the tables and concentrate on people rather than on statistics. Those who campaign to save research on animals (and the European Commission has been infiltrated by activists who want to stop it) once used graphs. Nowadays they have pictures of 1950s hospitals with children in iron lungs who would be saved today with a polio vaccine tested on chimpanzees.

The problems of distrust on one side balanced by secrecy on the other remain. On the other hand, today's suspicion is much healthier than the attitude of those pre-Brussels days when the Boffin Always Knew Best.

The Name of the Rose

Medical textbooks describe a condition called anomia: the inability to name everyday objects. One form involves the names of vegetables. Those afflicted go into grocers' shops with dread. They are forced to point at, or to ask for, those long orange things with green tops that go well with boiled beef.

Supermarkets and tinned carrots have put a stop to their problem, but a more widespread kind of anomia involves the names of people. I have, as it happens, for many years shared a mortgage with a woman called (as far as I remember) Norma Percy. In the difficult moment of first introduction of companion to parent I told my mother, with great confidence, that she was about to meet '–'. My mind went blank.

Students, too, are insulted when I cannot remember what they are called (although a stab at Hannah or Simon often does the job). I know exactly who they are – it's just the title I cannot come up with. The nifty device that identifies faces from a computer memory bank and displays the name on a pair of spectacles could help, but I doubt that I will buy one. The problem is buried deep in the brain.

Anomia involves a difficulty in making a connection between meaning (dim undergraduate) and sound (Si-mon). Studies of people with brains damaged by strokes or tumours show that the inability to remember the terms for particular classes of objects – vegetables, animals, tools or people – is associated with destruction of small parts of the temporal lobe on the left side. Brain scans show that normal people, as they think of lentils or marrows, camels or skunks, screwdrivers or drills, Einstein or the Queen (insofar as they do), activate groups of nerve cells in just the areas damaged in patients who are unable to remember what such things are called.

Those who find it hard to name tools (the one that cuts wood, or bangs in nails) have the same problem with animals (that thing that gives milk and goes mad, or the one that makes wool). However difficult it might be to find the word 'hammer' or 'cow', though, they manage just fine with people (the Prime Minister or the President of the United States). Once again a scan shows an overlap between the nerve cells that light up when tools or animals are labelled. The brain cells involved in the recognition of individual people are in a different place, a short distance away.

'Animal', 'vegetable', 'tool' or even 'student' all share one thing. Each is a concept: within each are tens of thousands of specific cases – elephants, peas, power drills, Cleopatra – as different as can be imagined, but each a member of a definable class that unites them as beast, plant, implement or human being.

Such objects can, though, be classified in different ways. Peas, pill-bugs, ball bearings and Billy Bunter are all (more or less) spherical and Cleopatra and an instrument-maker's plane are each more beautiful than Caliban or a lump-hammer, although most people would not immediately class them together on the grounds of shape or aesthetic attraction. However much we might argue about the details, and however much modern art or Britpop has muddied the waters, most people can identify round as distinct from square, music from noise, beauty from hideousness. The brain has, it seems, a series of centres able to sort things into categories in several different ways.

But how wide can those categories be? Philosophers once argued about whether beauty has a reality distinct from objects that are beautiful, or whether the idea is just a convention of language. Theology was involved, too. Does the concept of the Holy Trinity refer to three distinct Gods, or just one with three attributes? The ancient heresy called Sabellianism tended towards the latter: there was a single God with three talents, just as the Sun is hot, round and bright. The idea was quashed, with what rigour was necessary.

Some neuroscientists dismiss most brain-scan work as no more than molecular theology: an attempt to categorise the vastly complex, the brain, into a series of childishly simple, and largely arbitrary, categories. Even so, it does seem that the tool- or vegetable-recognition centres have a real identity; and that, at least so far as they are recognised by specific parts of the brain, such groups – be they tools, vegetables or gods – have a definable existence of their own. Could concepts as broad as justice, or good and evil, be coded for by a few nerve cells? And what will the theologians say about that?

An Unsteady Earth

I once stood up in a canoe. After my first trip in an ocean kayak, on the choppy waters of San Francisco Bay, I was so relieved to get closer to dry land and further from the wish to be sick that I rose to my feet a few yards from shore, with predictable (and dampening) results.

The problem came from an unexpected wobble. The balancing mechanism in the inner ear is pretty good at keeping us upright, but if the ground beneath its owner's feet shifts it cannot cope and vertical soon turns into horizontal. The same thing happened on the Millennium Bridge when it opened, because when hundreds of people quiver together their ears panic in synchrony and the sway grows until something (the bridge's own inertia in this case) stops it. They, too, reached the shore with a certain relief.

Such relief may be misplaced, for the land itself is far from stable. Look long enough at a distant star and it seems to circle around its true path. The apparent motion is smaller than the long-term unease of the Earth's axis caused by shifts in the movements of the Moon – nutation, as the process is called (to be distinguished from natation, which is what happens when you stand up in a canoe). The tremor is not the result of some remote commotion in the heavens but is closer to home.

Its cause was discovered by the amateur astronomer Seth Carlo Chandler in the 1890s. Like a top as it spins, the axis upon which the Earth turns tends to meander – at two inches a day and with a period of once in fourteen months – around the geographical poles. The diameter of this

Chandler Wobble varies from around ten to fifty feet across and changes in an unpredictable way from cycle to cycle. The world went through a major attack of the wobblies in 1910 but has settled down since.

Atlas, it seems, has shrugged (or swayed) – but why? To move the Earth takes energy and, like a gyroscope (or the Millennium Bridge), a spinning planet has a strong desire to stay upright (or straight). Did some ancient bang push the world off centre? If so, the evidence should be in the Earth's crust, but geologists find no sign of it. Something keeps our home in its tipsy state – but what?

Perhaps the global restlessness comes from the drift of the continents (and some move at a foot a year), or from the giant sections of ice that float away from Antarctica. Perhaps the Earth is pushed off balance by volcanoes, or an intestinal eruption deep within the core or the population of China rushing from East to West in response to an imperial command. All (or most) of these ideas are feasible, but we have no firm evidence for any.

A decade-long comparison of the size of the Chandler Wobble with computer models of the atmosphere shows that in fact an apparently tiny force makes the world sway. The wind is blowing our planet about. It does most of its work indirectly, in great storms that move the sea and generate waves. As the ocean surges to and fro like water on the car deck of the *Herald of Free Enterprise* (the ferry that sank off Zeebrugge because its bow doors were open) it causes the globe to tip and its axis to move.

The experience that bothered my inner ear beneath the Golden Gate hence has a tie with the restive but undetectable movements of the less than solid Earth; but why should anyone care? What, on a scale of tens of thousands of miles, does an annual shift of a few feet matter? Today's maps are so accurate that the wobble makes a real difference. Trident submarines use a system called StarTracker to guide their missiles to a target. If the Chandler Wobble were not programmed in, a missile fired at, say, Tehran from the seas off San Francisco could miss by as much as a mile – which may not seem a lot but, with warheads directed at a hardened military structure, is important. The Global Positioning System also has to cope with the Earth's flighty behaviour and its computers must be set to compensate for annual shifts in its response to the wind.

It was embarrassing to stand soaked and sea-sick upon the breezy shores of Sausalito, but more so to be forced to decline the offer of a bike to go off and get dry clothes. So uncertain is my own internal balance that I have never learned to control that particular cyclic wobble and I had to drive instead.

On Sunbathing Snails

According to Ron Glum in the classic radio series that bears his name, summer days are long because they expand in the heat. Real physics is not quite as simple, but the fit between day-length and temperature is easy enough to understand. The Earth is tilted on its axis, and as the year wears on and our planet moves around the Sun, our nearby star seems to rise and fall in the sky. Its high point in the Northern Hemisphere comes when the North Pole is tilted towards it. That makes midsummer's day longer than any other.

How much solar energy will hit the ground depends on several things. In June the source of heat rises high in the sky and its rays have to pass through a thinner slice of the atmosphere than in winter, when more of its powers are soaked up by air and dust. A shortage of clouds also allows extra sunlight to stream in.

The main effect on input is more subtle and depends on the angle the Sun's rays make with the Earth's surface. People who warm themselves in front of a fire find that it pays to sit in front of it and not off at the side. The same is true of those who bask at the solar furnace. At midsummer we have as direct a line as possible to our energy source, which means that its heat is concentrated on to a smaller area than at any other time.

Away from the equator, the effect is large. At noon on 21 June in London, a vertical metre stick casts a shadow about half a metre long. On 21 December (assuming it casts a shadow at all) it will stretch for six times as far, as proof of how the Sun's energies are dissipated in the depths of winter. The same happens as each day wears on – long shadows (and not much solar input) at dawn and dusk, shorter ones (and more calories per square metre) at midday.

Farmers are well aware of the importance of light and shade, and some have turned to mixed crops of tall and short plants in which the energy that passes through the canopy of the tall partner benefits the low-growing variety below. In the science of shadows people try to measure how light penetrates various kinds of vegetation at different times of year and of day. They use electronic scanners, or turn to satellites to see how much of the Sun's output bounces back into space from the field, or forest, being tested.

Many small creatures depend for life and death on their ability to stay within a narrow range of temperature. They shuttle in and out of the sun as the seasons, and the hours, go on. Snails do not attract much cash for scanners or satellites; so in an attempt to see how they cope with light and shade I once defined myself – in the tradition of Louis XIV, *le Roi Soleil* – as the midsummer Sun itself.

How does a snail on a rocky slope, or in a thick hedge, experience the passage of its energy source across the sky? To find out is easy enough. Take twenty plastic snail-sized spheres and lob them into the vegetation. How they fall depends on the habitat: a thorn hedge stops most of them before they reach the ground, while a patch of grass allows the majority to tumble straight through on to the ground.

A snail looks at the Sun through a filter of vegetation. The Sun looks at the snail in just the same way. To test the animal's exposure to the fount of all power, an observer needs only to follow the azimuth – the track of the Sun across the sky – and to count the number of scattered spheres visible from twenty or so different positions at intervals from sunrise to sunset. A stepladder does the job cheaper than a satellite; and to avoid complicated calculations, Apollo's chariot is tracked across the heavens as if every day is Midsummer's Day, whatever the actual date.

A snail's-eye view of the solar system comes up with an interesting result. Places with shifting shadows and sunflecks have a more diverse molluscan population than do those that are simpler – always gloomy or always exposed. Why, we do not understand, but local variation in solar input may allow each animal to choose the microclimate it prefers – to sunbathe or to creep away.

Like a sundial compared to a Swatch, Jones' balls, as they are known, are crude but effective. One problem comes on windy days, when the spheres get knocked to the ground. As Ron Glum might explain, that is because the trees lash their branches about and whip up a breeze.

Proud to Be British

As all politicians know, nationalism always returns to the past. Britain harps less on the glorious days of Empire than once it did, but the avalanche of films about the Normandy Landings, the Desert War and all the rest continues without respite, sixty years on.

Even fossils have been hijacked to give a nation an identity that stretches back before history began. When Boxgrove Man was discovered in Sussex in 1994, the chief archaeologist for English Heritage (which manages the site) said, bizarrely: 'Now every Englishman can stand a little taller!' Somehow a scientific specimen had been transformed into a symbol of patriotic pride – which, foolish as the idea is, at least helped a few bone fragments to make the national news and to generate excited speculation about what our national habits may have been half a million years ago. Soon afterwards a bone from Atapuerca in Spain was found to be three hundred thousand years older and it was the Spaniards' turn to dance in the streets. The Brits have managed to push their history back to within a hundred thousand years of those remains with new material from East Anglia, but as the bones were not record-breakers nobody took much notice.

Prehistoric nationalism goes back a long way. The founder of modern archaeology was a Frenchman with the glorious name of Boucher de Crèvecour de Perthes – the Heart-breaking Butcher of Perthes. In 1837 he found some oddly shaped stones made, he thought, by ancient people. At first the idea was mocked, but the stones were accepted as relics of ancient times as soon the French realised that they showed the origin of civilisation to be in France, still its capital. One enthusiast saw an even nobler lesson in Gallic history. To him a skeleton found in the Dordogne looked like those of modern Asians and one from the Riviera resembled those of Africans. Cro-Magnon Man (the great cave painter, himself named after a village in the Dordogne) was similar to today's Europeans. The result was gratifying and clear. France was the birthplace not just of civilisation but of all the peoples of the world.

The Germans had a rather different view. They pointed instead at the wonders of Neanderthal Man (named after his home in the Neander Valley, near Düsseldorf). Tall, good-looking in an Aryan sort of way – and with prominent brow ridges, developed from his habit of frowning while deep in Teutonic thought; a true Kant of ancient times. The British were just as chauvinistic but with less excuse. Piltdown Man, alleged to be the first Englishman, was a clumsy fraud, but a fraud with a national identity, for he was found with a bone implement said to look like a cricket bat.

The furore about fossil evidence, from Cro-Magnon to Boxgrove, brings to mind another English hero, Sir John Falstaff. Prince Hal criticised the bloated knight's bar bill with: 'O monstrous! But one half pennyworth of bread to this intolerable deal of sack!' Many palaeontologists suffer from Falstaff Syndrome, a deficiency disease in which an ounce of bones is washed down with a barrel of speculation.

Boxgrove Man was in fact just Boxgrove Bits; a piece of shinbone and a couple of teeth, together with some stone tools and the antler used to make them. Cut marks on the remains of a rhinoceros found with those relics show Mr

Boxgrove to have been right-handed. From those modest clues, scientists have constructed a remarkably complete story of English prehistoric life.

Because he carried an antler to shape flints, Boxgrove could, they claim, plan ahead. He was a meat eater and hence probably had a large brain, for, say the fossil hunters, one follows from the other. Brains are expensive; so are guts. You can have lots of either of them, but not both. The Old Man of Sussex solved his intestinal problem with a shift to a high-quality diet – meat – and went easy on the bowels. The key to the British intellect lies, it seems, in the alimentary canal.

Such a discovery is best celebrated in a butt of sack – but the few fragments of Boxgrove Man do bear some kind of political message. He was not a citizen of these islands – for they were not islands at the time, but attached to the Continent. Even his name, *Homo Heidelbergensis*, has a German root. To make a guess at what the creature actually looked like, scientists had to fit the Sussex shinbone with a skull from Greece, a jaw from Germany and a French pelvis. Boxgrove Man was, without doubt, the first European.

Sharks, Soap Bubbles and Toxic Dumps

Nature often imitates *Art Monthly*, for the world's prime scientific journal has a regular column on science and the arts. For most scientists the two ways of thought seem to have rather little to say to each other. Leonardo da Vinci tried (not to speak of Damien Hirst), but most modern attempts to unite art and science generate the deep indifference intrinsic to a pickled shark. To an artist such an object may look like science, to a scientist like a work of art, but in reality it is no more than a cheap publicity trick.

Some of *Nature*'s artistic offerings have themselves been somewhat eccentric (navel lint from a deep-sea fisherman was once the featured work). Others, though, are full of fascination.

A series of sculptures by Jonathan Cullen was made by allowing cement powder to pour through holes drilled at random in a plate, leaving a set of pits behind. Viewed from above, the cavities give a sense of flying over the Alps, with peaks and ridges scattered below. Handsome enough, but not, it seems, very scientific.

The pattern left when powder pours through random holes emerges because each grain is bound to fall through the nearest opening. The ridges that build up between the pits hence mark the edges of each hole's catchment area. In turn this forms a geometrical structure called a Voronoi cell, with each cell bounded by the lines of equal distance from its nearest set of holes. The boundary ridges are easy to draw: join a hole to its nearest neighbour by a straight line and draw a vertical halfway along it. Where the verticals intersect gives the edges of the cells.

The arrangement looks rather like the pieces in a stained-glass window. It also bears a remarkable resemblance to the structure of the universe. Galaxies are not scattered at random through the vastness of space. Instead they lie in thin sheets, separated by huge distances from the next galactic layer. Their pattern follows the boundaries of Voronoi cells (the mathematics is made more difficult because they are sheets in three dimensions and not lines on a flat plate).

Closer to home, soap bubbles form a Voronoi foam in which the shape of each bubble follows the rules and maintains a fixed volume with a minimum area of liquid. Venus is often painted emerging from the foam, to give another tie between astronomy and the arts (although a hard look at the Botticelli version shows no sign of the famous cells in the white froth beneath her oyster shell).

The parallel between science and society goes further. To map the cells around Iron Age forts in Britain estimates the sphere of influence of each local chieftain. In modern times Voronoi maps are used to design the best network of roads to serve a set of scattered buildings. They also solve what geometers call the 'toxic dump' problem. Given a set of villages scattered over the landscape, what is the largest empty circle that can be squeezed between them? Voronoi gives the answer; and the point where best to put a rubbish tip, a nuclear power station or an exhibition of pickled sharks.

The pattern is also found in the living world. Territorial animals have to pack themselves in to what space is available. The pattern is a set of the famous cells, their edges delineating each territory. Male red grouse anxious to find a mate search out and hold their own piece of land. Shoot a male and another moves into his space, proof of how crowded life must be. If a bird is injected with the male hormone testosterone the size of his patch increases, for like a bodybuilder attacked by steroid rage he displaces his unlucky neighbours from their homes. The ripples from the obnoxious male spread across the landscape. Many cells have to rearrange themselves until some unfortunate bird a long way from the action finds himself squeezed out; elegant mathematics no consolation for sexual failure.

Dr Johnson made an inadvertent comment on the Voronoi cell. He toyed with the idea of a visit to the Giant's Causeway, the famous Ulster landmark made of columns of basalt. The shapes of the pillars are just as predicted by theory, for as they cooled and contracted they were forced to split – like a white-hot foam – into irregular polygons of rock that minimised the enormous strain. To Dr Johnson, the causeway was 'Worth seeing, yes; but not worth going to see'. Britart, in fact, exactly.

On Fat, Age and Sex

For an academic, the definition of middle age is when students stop looking younger every year. The first week of the new session is an unwelcome reminder of that fact. To me the new entrants look, as they have for the past decade, just like last year's crop; and I find it hard to remember the days when I was just as juvenile myself.

Students may be ageless, but each year they get bigger. A line in Dylan Thomas is a great help to stunted Celts (five foot seven in my case): 'he was of middling height – middling for Wales, that is'. One-point-seven-o metres (as my passport defines it) in the 1960s allowed me to tower modestly over at least some of my female fellows. Nowadays, though, I am looked down on by almost all the first-year class, whatever their sex (and even before they have heard any lectures). Teenagers are taller than they were, with an average increase of three inches in the past forty years. Europeans have become loftier and Americans fatter (and we equivocal Brits are both: second tallest after the Dutch – giants at a mean male height of six feet – but the heaviest in Europe).

Such a change is, in biological terms, spectacular. After all, humankind has shrunk, not grown, in the one hundred and fifty millennia since the first members of our species did their A-levels. At the present rate we will outgrow those early *Homo sapiens* by the end of the century. The environment, rather than the genes, is to blame. Our national diet has changed; and although nobody is certain quite what was lacking in the 1950s, the shift has had dramatic effects.

Height is a proxy for well-being. In the eighteenth century the richest men were three inches taller than the poorest, while now the difference is just over an inch – which is still quite a lot.

Such shifts have happened before. At the origin of agriculture, ten thousand years ago, the average dropped by three inches as people shifted from the healthy diet of the hunter-gatherers to a degenerate existence based on a primitive form of muesli. The late twentieth century was a time of protein and fat and its adolescents shot up in response.

Today's students are also more mature than I was. Not, as a visit to the college bar shows, in their behaviour (hard to believe, but they are even drunker than my generation) but in their development. The age of puberty is in rapid decline. In the 1830s, the records show, most girls in Scandinavia reached menarche at the age of seventeen, whereas nowadays they attain the mature state four years earlier. Boys, too, have a shorter childhood than before.

Food and sex go together. What sets the time of puberty is not age itself but body mass; and the weight (rather than the birthday) at which teenagers become mature has not altered at all in the past century and a half. Today's children grow faster, reach the critical size at a younger age and, as a result, become adult earlier.

It all has to do with those turbulent hormones. In adults the hypothalamus – the brain's chemical control centre – makes a pulse of a protein called luteinising hormone every few minutes. This regulates both male and female sexual function, and is one of the switches for puberty. Luteinising hormone responds to food and body weight, with an increase in output after a good meal and a drop in times of starvation or exercise (which is why anorexic women and female athletes may suppress their menstrual cycle). In children the hormone responds to the increase in size, and helps puberty to kick in earlier.

The chemical is just one member of a great family of hormones that link sex with fat. Not long ago a protein called leptin was found to be involved in the control of appetite. A new and spherical mouse mutant appeared in a laboratory stock. It had problems with its leptin system and grew until it could grow no more. The hormone is made by fat cells and tells the body how much energy it has in reserve. If there is too little, the order to eat goes out. Fat itself is an endocrine organ and fat people have much more of this sinister substance than do those of normal weight.

When the chemical is injected into young mice they come to sexual maturity within a few days, giving another tie of size with puberty. The two states are linked in other ways. Mice lacking the cell receptor to which insulin sticks are hungry, thin and sterile. As the cheeseburger generation drugs its way to middle-aged slimness and – with luck – to rejuvenated sex, the new anti-appetite pill may turn out to be a contraceptive as well.

I Got There First!

Britain was once obsessed with the vexed issue of precedence. On formal occasions the Sovereign is always in front, followed by her Consort, the Lord Privy Seal, the Master of the Horse and a gaggle of Dukes and Earls, with the Younger Sons of Knights trailing in the rear (apart, of course, from all the women except one).

Scientists are just as bad, although they have a certain excuse for touchiness as their careers depend on being the first rather than the second to come up with a particular discovery. For the aristocrats of science, though, simple egotism is just as much to blame.

In 2003 the Nobel Prize for Physiology or Medicine was shared by the American Paul Lauterbur and the Briton Sir Peter Mansfield, for their invention of magnetic resonance imaging, a technique based on the use of magnetic fields to look inside the living body. The award was greeted with outrage by a fellow physicist, Raymond Damadian. He insisted that he had the idea first and deserved a share. Damadian was without doubt much involved in the early days of the work and to press his Nobel claim took out full-page newspaper ads in which he described his omission as 'a shameful wrong that must be righted' (although in a later fit of pique he maintained that he no longer wanted the 'corrupt' prize, which had 'the stain of Cain upon it'). The Nobel Committee was unimpressed and stayed silent.

Very regrettable, very ungentlemanly and very unBritish, we might, with a certain smugness, conclude. Not so, for when egos are bruised our own scientists are just as bad. Newton was infuriated by the assertion that calculus was invented not by him but by the perfidious Saxon Leibniz. In fact the two did the job independently, but Sir Isaac went to great lengths to undermine his rival and packed a Royal Society inquiry with friends to prove that Leibniz had stolen the idea. He then took revenge on Robert Hooke over the latter's work on gravity and optics (priority again) and ended his life with a persecution mania (fortunately, neither newspapers nor Nobels had been invented at the time).

Charles Darwin, in contrast, was a gent. When Alfred Russel Wallace wrote from the Far East with his independent idea about evolution, Darwin at once conceded joint authorship (and Wallace reciprocated with a book called *Darwinism*). In the 'Historical Sketch' which prefaces later editions of *The Origin of Species* he allows himself to be testy about a few of the many who had claimed priority ('As Dr Freke has now published his Essay . . . the difficult attempt to give any idea of his views would be superfluous on my part') but he accepted the claims of others who had come up with parts of the theory of evolution before he did.

The community of scholars that picks lint from Darwin's navel has chanced upon a hint that the great man may owe more to the past than he admitted. In 1794, in his two-thousand-page *An Investigation of the Principles of Knowledge*, the geologist James Hutton penned a crucial paragraph: 'those which depart most from the best adapted constitution will be most liable to perish, while those . . . which most approach to the best constitution for the present circumstances will be best adapted to continue, in preserving themselves and multiplying the individuals of their race' – a clear statement of natural selection, sixty-five years before *The Origin*.

Hutton belonged to the Edinburgh scientific community which influenced Darwin throughout his life. Perhaps the idea of evolution was planted in the great man's mind in his days in Scotland, to surface half a century later. Had he realised, Darwin would no doubt have acknowledged the truth.

Raymond Damadian, like Newton, is a fundamentalist Christian (and some blame his Nobel rejection on his religious views). He denies Darwinism, for life, he says, began with a miracle just six thousand years ago (not much chance of a second bite at the prize there). I beg to differ about the origin of life, and while on the subject of precedence have a certain claim of my own.

A whole evolutionary industry now works out the size of ancient populations from the genes of their descendants. Its most publicised finding is that *Homo sapiens* was once an endangered species, with just a few thousand of us around in Africa, and that we went through a bottleneck as we left our native continent. Let me state my case as modestly as I can: I DID IT FIRST! (albeit with the help of a clever Iranian physicist, Shahin Rouhani). There it is; the journal *Nature* for 1986, volume 319, page 449 – 'Human evolution: How small was the bottleneck?' (By comparing patterns of variation in a certain gene in Africa versus the rest of the world, we showed that no more than about eighty people made the exit.)

Did anyone take any notice? No. Is the paper ever referred to in the literature? No again. Has it improved my (modest) scientific reputation? Certainly not. Do I care? No, honestly, hand on heart, not in the slightest; it means nothing to me, I'm not bitter, I don't mind at all. I do, though, regret wasting so much time learning Swedish.

Dining Down the Food Chain

There is no accounting for tastes. Radio presenters always need a few words to set the sound level before they start and often ask their contributors what they had for breakfast. Over the years the typical answer has changed from bacon and eggs to muesli. One interviewee claimed to have eaten whale on toast, but he was joking (or perhaps it was a very small whale).

Certain whales have had a similar shift in diet. On the coast of Alaska some killer whales have moved from a breakfast based on seals to one whose main ingredient is sea otters. The change in tastes is, needless to say, bad news from the prey's point of view, as to go to work on an otter, toothsome as it may be, is not enough; one whale needs half a dozen of those furry snacks each day – two thousand a year – should it wish to stoke up on that item alone. The whales' change in habits has led to an otter collapse and the numbers of those beasts are only a tenth of what they were thirty years ago.

For the sea otter, as for so many other creatures, the great chain of ecological being has been broken by human stupidity. First, huge numbers of fish were taken to feed hungry Americans and the seals were starved out. The killer whales then had to find another source of protein and turned to otters. Along five hundred miles of Alaskan coast they eat eight thousand of them each year. A change of diet by a mere four killer whales was, at least in principle, enough to destroy the sea otter.

The hungry leviathans' new habits had an unexpected

side-effect. The huge kelp forests of Alaska, home to thousands of fish, are on the way out. The otters ate sea urchins, which ate kelp. Today the coast has fewer otters than before, many more urchins, and in some places the kelp has almost gone. The addition of a single link at the top of the food chain – perhaps just one otter-eating whale in a hundred miles of coast – had a huge effect on the plant at the bottom.

The fate of the kelp forest is a reminder of how short nature's game of consequences can be. Most ecosystems have no more than three or four steps from bottom to top: plankton to fish to seal to whale; or kelp to urchin to otter. Add a link, or take one away, and the effects are drastic.

Why are food chains so short? It's down to efficiency. When an urchin eats seaweed it gains no more than a tenth of its goodness. The otter does rather better when it swallows the urchin and the whale better still but, even so, each one wastes more than half the worth of its prey. For an insect-eater (or a devotee of larks' tongues) just one part in a hundred of the value of each item may be usable. As a result, in nature (as in business) too many middlemen leave too little for the final consumer and it becomes impossible on economic grounds to add another link. Evolution will never come up with a beast that preys on killer whales, for one individual would need a whole ocean to support itself (which is why the whaling industry went bust).

Experiments on bacteria in the laboratory and on frogs in tropical tree hollows filled with water show that when resources are added in the form of nutriments or light, there can be more steps from bottom to top; but rich as the raw material might be, the ecological ladder almost never has more than five rungs.

We humans see ourselves as being outside such rules. With computers and cars, we are right, for there may be dozens of steps between crude oil and plastic keyboard. With food, though, nature is less indulgent to human whims. Not long ago, like sea otters among salmon, we could afford to hunt the aristocrats of the piscine world. Tuna and swordfish live four or five steps from the bottom of their food chains (which put their landbound consumers at levels five or six). A diet based on top predators will not last and those large and elegant animals are on the way out. In the last two decades the world's fishermen have shifted, on average, a whole step down in the food chain in the search for prey. In the South Pacific some have turned to krill – the tiny crustacean eaten by the great whales who filter water; just one rung up from the plankton who soak up the Sun's rays and use the energy to feed on the ocean's chemicals.

The krill fishermen sell their catch to fish-meal factories who make it into raw material to feed the chickens or pigs which feed us. The food chain has five steps when the plankton are included – but it needs lots of extras in terms of transport and machinery to keep it in business. Cheap oil is the fuel of complex societies, and to add an extra layer between sea and supermarket demands even more energy, itself a limited resource.

How long will it be before the first radio interviewee admits to eating plankton for breakfast?

Give Me a Call Some Time

The modern world is filled with irritations, much exercised by newspaper columnists and by television series about grumpy old men and women. Such people are predictably peevish about some rather obvious annoyances. However, old men forget; and one forgotten irritation of the modern world is the mess that has been made of our telephone numbers.

My home number (with some creative editing) is 7881 1982, unless I call from far away, when it becomes 0207 881 1982. When I lived in Edinburgh it was 229 4561 within the city but 0131 229 4561 from outside. Should I be fortunate enough to be snapped up by one of the older provincial universities – Cambridge, say – I would not have eight digits for local and eleven for long-distance calls (London), or seven and eleven (Edinburgh), but six and eleven.

Compare our fractured system to the simplicity of France, where all numbers, dialled from anywhere, consist of ten figures, always grouped in twos. Our numbers are a typical British muddle, a fudge at the expense of the poor customer, whose memory as a result suffers from digital overload.

Most people cannot recall more than about seven figures in a row when they are asked to recite back a string of digits just after hearing them (which is vexatious when trying to make a long-distance call). The information disappears almost at once. What was my Edinburgh number again? Yes, the delay since reading the second paragraph of this piece has wiped it out, although I could summon it up forty years on.

I can do so because I have moved the data from one memory bank to another – from short-term to long-term. The two accounts have different investment rules. To make a long-term deposit (and in the brain long-term means more than a minute or so) is hard work. It involves the synthesis of new proteins. The genes responsible are related to those involved in DNA repair, which is unexpected. Short-term memory, in contrast, relies on cheap but unstable signal chemicals that soon fade away.

The move from short to long involves, just like getting to Carnegie Hall, practice, practice and practice. Some people are prodigies. Hiroyuko Goto of Japan holds the world record for the recitation of pi – the ratio of the circumference of a circle to its diameter – by chanting 42,195 digits with no mistakes. To win the prize meant plenty of rote learning and a great deal of burdensome protein synthesis.

But when it comes to the immediate recital of a hail of figures, nobody – try as they might – can get much above ten in a row. A tiny spot in the base of the brain lights up when people are given a salvo of new information, and there are great individual differences in how big it is (whether Cambridge people, with their reduced telephone code, have a smaller version than do Londoners is not yet clear).

For those who fail the test several tricks can help. People are better at language than at sums, and to repeat a telephone number under one's breath keeps it in play for longer. Some kind of association with a memory already stored away is also useful, and it is harder to forget the number 999 1066 than a random example culled from the Yellow Pages. Pattern makes a useful aide-memoire, too: 2468 is easier than 7219. Most of all, 'chunking' – sticking blocks of data together to form a smaller set – helps us to keep information in our heads.

Electronic directories give out each separate figure in a deadpan way almost designed to stop the listener from using such a trick. A line of single figures is far more difficult to retain than are groups separated by the vocal brackets used by human operators – which leads to another absurdity in the British set-up. The French quote their telephone number as she is spoken – zero-four, sixty-seven, ninety-six, twenty-one, thirty-eight, which is easier to retain than 0 4 6 7 9 6 2 1 3 8. Our barmy system, with different places using different patterns, makes such shortcuts impossible.

The whole idea of distinct local, long-distance or national systems will soon be out of date. The companies are moving over to 'packet switching' – sending bursts of data over the internet – and to do this will all have to use the same codes. The world has fewer than a billion phones, and with ten digits each one could be given a unique number. I have put in an early bid for 12 34 56 78 90.

A Stiff Upper Lip for Africa

Those who believe in the trickle-down theory of economics – that if you feed the donkey enough grain the chickens will grow fat on its dung – have something new to celebrate. It has emerged from a cure for sleeping sickness, a tropical disease that infects hundreds of thousands of people a year and kills a good number of them. The illness is caused by a blood parasite transmitted by tsetse flies. Once the battle against it seemed almost won, but after many years in remission the disease is back, and in many villages in the Congo, whose health system has been destroyed by civil unrest, half the population have died of it.

A treatment about to be abandoned on the grounds that it did not make enough money has gained new life from an unexpected source. The concern of American ladies about their unwanted moustache may save the lives of thousands of African children.

The parasite in question is a trypanosome that gets into the blood and reaches the brain, leading to coma and to death. Like the malaria parasite, it confuses the host's immune system with rapid changes of identity in the first few days after infection, which frustrates all attempts to develop a vaccine against it. However, the miracle drug eflornithine kills the agent of disease by interfering with its ability to synthesis DNA. In most cases it effects a complete cure.

Treatment costs around seventy dollars a head and, as so often, those most in need cannot afford it. Eflornithine was first developed as a possible specific against cancer and until the late 1990s industry continued to produce it in the hope that some day they would make a (metaphorical) killing. Their hopes were dashed in a clinical trial which showed no real effect on the disease, the factories closed down and, for African kids with sleeping sickness, the prospects became grim.

Then eflornithine was found – out of the blue – to have an unexpected talent, for it prevents the growth of facial hair. As a result, a huge market has opened up. The Gillette company estimates that forty million American women suffer from that problem and might buy the solution. A cream aimed at women has been approved in the United States and, at two dollars a gram, promised vast profits. The production lines for the wonder drug, under its new name of Vaniqa (seen as friendlier than its medical name, Ornidil), have gone back into action. As a result, the diseased chickens of the Third World will profit from the beauteous donkeys of the First as the treatment trickles into Africa (although that is too cynical a view, for the companies who make it promise to give the medicine away for three years and will then sell the drug at cost).

Eflornithine switches off the enzyme that makes facial hair flourish. An antidote might help the moustached ladies' partners, many of whom face the opposite problem. The Bald Headed Men of America, from their headquarters in Morehead City, North Carolina, estimate that a billion and a half dollars a year is spent on hair replacement. Several drugs have unexpected side-effects on the top of the head. Minoxidil is used to treat heart problems but can also slow the tide of scalp. Finasteride, an inhibitor of the enzyme that converts testosterone into its more potent form, was developed to treat enlarged prostates (rather than fore-heads). In lower doses it, too, has been a help to those with follicle problems.

The myth that bald men have extra sexual power has no foundation (although Romans did chant politely to Julius Caesar: 'Guard your wives: the bald adulterer comes!'). Politicians find it hard to know what to do about their crowning glory. Most remain clean-shaven; and, in Russia, Peter the Great detested facial hair so much that he shaved his nobles himself. Whatever his prejudices, for much of history beards (if not baldness) have been a symbol of authority. In ancient Egypt, indeed, Queen Matshrpdont wore a false one, gilded and filled with jewels to show her political importance.

To Hillary Clinton, if she wishes to be President, the message is clear: she should lay off the eflornithine, just in case. She might also remember that enough of the drug to treat everyone infected with sleeping sickness could be bought for just twenty million dollars a year, a tiny fraction of the amount spent by men on attempts to rescue their scalps. And, in a positive spin on the trickle-down tale, the Bill and Melinda Gates Foundation has poured in cash to find new treatments for sleeping sickness. With luck, they may even find a baldness cure on the way.

O Ye Daughters of Jerusalem

Back from a trip to Jerusalem (the place, not the oldest pub in England) I am impressed by the differences among intellectual maps of the world. The city contains three fanaticisms, most obvious in the response of those who share them to those who do not. After abuse from an Islamic guard at the Dome of the Rock there was rudeness from a gloomy Armenian in the Church of the Holy Sepulchre. Finally, in the Orthodox quarter, a brick bounced off the car just because I had the temerity to drive down the street. All my adversaries shared the same glassy-eyed certainty that they were right and the rest of mankind wrong; but, by definition, at least two must be mistaken.

Jerusalem has the misfortune to be at the heart of the universe for all three. The shops sell copies of ancient charts, all with the city in a central position with other nations around it, as a clear reflection of the beliefs of those who drew them.

Geographers still use 'mental maps' to test how people see the world. Just as in the Middle Ages, they ask a subject to sketch the position of his home in the context of the globe as a whole. The results are illuminating – and depressing. A map of a neighbourhood, a city or a country drawn by someone on the breadline is very different from the same thing produced by a member of the middle classes.

Much of the work was done in Chicago. The universe as drawn by a ghetto dweller consisted of just a few streets, with a link to the local hospital and perhaps to the courthouse. Almost always the subject's own home was at the centre of the plan. Beyond that, all was vague. New York and California might be in about the right place, but the poor could not draw a map of their nation as a whole because they had never experienced it. The rest of the world? – forget it. (I was myself once asked by an African-American lady whether I had met her brother, who was in the 'France–England area'.) The rich, in contrast, spent hours on personal cartography and could produce an exquisitely detailed sketch bearing hundreds of place names (and they did well even for France and England), with their own abode located to within a few hundred yards.

Mental maps sample people's views of the world they live in. But how many of us could draw a map which extends beyond such a narrow horizon to show where we stand in relation to the cosmos? I could not, but I speak as a man who was due on a live radio programme at 7 p.m. in Worthing and blithely got off the train with an hour to spare at Woking, as proof of how incomplete is my own map even of England (I made it with the help of a monstrous outlay on taxis). The Earth goes round the Sun, I know; but where the Sun lies in the universe I have almost no idea. We are somewhere at the edge of the galaxy in a rather suburban little cluster of stars but on the larger scale my mental map is a blank.

Astronomers now know where we are in relation to the rest of creation. The centre of the universal mental map is not Jerusalem, nor anywhere else within human experience. Instead, at the core of our galaxy is a massive black hole, hidden away behind the Milky Way.

A once obscure object, Sgr A*, is the hub around which our lives rotate. It has a mass a million times greater than the Sun and swallows most of its own emissions before they have a chance to escape. The discovery unites our own part of the universe with the many other galaxies that turn around an 'active galactic nucleus'. Those astral holy cities vary in kind from the sullen introspection of our own galactic core to extravagant displays of radiation called blazars, quasars or optically brilliant variables.

Such is the astronomical Jerusalem around which the firmament revolves; up in the heavens, out of sight of the naked eye. Oddly enough, the faithful of the medieval city believed much the same. There was, they thought, a facsimile of their sacred home hovering above its Earth-bound parent to which the chosen went when they died.

There was a dispute among savants. Jews put the New Jerusalem twenty miles overhead, Christians forty. Five hundred years on, the debate remains unresolved. Because the argument turns on faith alone, reason is of no use; and, as is always true in matters of certainty, science cannot help. In the Middle Ages the only solution was to throw bricks at the opposition. Some things, it seems, never change.

Choose Your Poison

Why are so many creatures full of noxious chemicals? And why are they so much more poisonous than seems necessary? After all, one gram of botulinus toxin is enough to exterminate thousands of people, which some might see as overkill. And why are natural poisons, drugs, so attractive to so many people – young men most of all?

Chemical abuse has undoubted pleasures (which is, after all, why we drink tea). The state is, in general, against hedonism and goes to great lengths to stamp it out. Queensland, for example, exerts severe penalties upon citizens found licking cane toads. As anyone who walks in an Australian forest soon finds out, most of its inhabitants are anxious to sting, bite or otherwise molest those who come near. Some new, vivid and probably illegal experiences await the first Australian macho enough to smoke a funnel-web spider or to make snake-venom tea.

The misuse of drugs might, though, be more than a search for simple pleasure. Instead it could be a signal in the battle of the sexes. After all, say some biologists, if a male is tough enough to withstand chemical self-abuse he must be hardy enough to be an effective mate. Any sensible female will fall for him. All those vile and emetic potions forced on young tribesmen as a rite of passage – or the drunken buffoonery at the rugby-club dance – is in fact a subtle hint of sexual prowess by those who can still stand up after a dozen pints, or puffs.

In the animal world, sex and venom often go together. A pretty Florida moth is full of bitter compounds designed to frighten off predators. As is true of many moths (and of all tea drinkers and most drug addicts), they get their preferred poison from plants. The female goes so far as to pass the chemicals on to her eggs, protecting them from attack.

The rugby-club theory can be tested by feeding the insects on plants free of poison, as to do so produces perfectly edible adults. Present those innocuous males to a potential partner and their virtuous upbringing at once removes all sexual allure. Given a choice, the females avoid those chaste bores and go for their natural, and toxic, cousins (which may be good news for boozers at least).

Females test how much toxin a male contains (and how sexy he is) by how he smells. Only those with a noxious odour are accepted. Selfishness plays a part, for a simple experiment shows that she gains an important nuptial gift from her other half. A wholesome female brought up on the poison-free food is made noxious after mating with a virulent partner, for she obtains the toxins from his sperm. The most poisonous females of all – those most *fatales* among the *femmes* – attain their elevated state by accepting sperm from many males, one after the other.

Other insects, such as the beetles that produce cantharidin, the main ingredient of the supposed aphrodisiac Spanish fly, do the same. The second-hand effects of that notorious poison are manifest in the tale of the Foreign Legionnaires in Morocco who suffered prolonged erections in spite of all they could do to cure them. Their drinks had been doctored by *Cantharides* beetles falling from the trees above their camp dining table. A certain flexibility about where they ate their meals solved the problem.

My single experience with deliberate self-poisoning came from a nervous attempt at fugu, the Japanese fish said to give the gourmet a delicious tingle as its neurotoxin gets to work. That too comes at second hand, for the fish picks it up from the tiny creatures that cause toxic red tides in the ocean. The protein interferes with the channels in cell membranes that serve to pump sodium and can kill within a few minutes. It is concentrated in the fish's liver, skin and eggs, which must be dissected out before the meal is served. The job involves much ceremony and restaurants are licensed to ensure that their chefs do not bump off too many customers. Even so, up to fifty deaths a year were once reported (which meant that it was a very apprehensive try on my part). Sad to say, I got absolutely none of the sense of 'floating on air' described by enthusiasts and in fact took no risks. Too late I learned that most of today's fugu are quite safe, for they are grown in hatcheries away from the source of the toxin.

Are women still impressed by a night out in a fugu restaurant – and by men who abuse alcohol-free lager and decaffeinated coffee? Or does sex always call for fatal attraction?

A Biologist's Bikini

The world's greatest biologist once wrote, with a certain humility, that: 'I have deeply regretted that I did not proceed far enough at least to understand something of the great leading principles of mathematics; for men thus endowed seem to have an extra sense. But I do not believe that I should ever have succeeded beyond a very low grade.' Thus Charles Darwin, admitting his own inability to do sums.

Whatever his weaknesses, Darwin changed biology for ever. His clever cousin Francis Galton was more confident about arithmetic ('wherever you can: count') but achieved far less. He counted many odd – and eccentric – things, to little end. Nowadays only archbishops have much interest in his paper on 'Statistical Tests for the Efficacy of Prayer' and writers of cookbooks in his famous work 'On Cutting a Round Cake on Scientific Principles'.

Galton was a clever dilettante who used mathematics for amusement, but Darwin, innumerate as he may have been, was a revolutionary. Although he managed without maths he might have been even greater with its help; but he was better engaged in riding across the pampas, watching a dog smile and, most of all, in just thinking. His own subject – evolution – became a haunt of mathematical biologists who constructed an edifice of austere beauty that was, alas, almost useless until grubby toilers in the scientific vineyard came up with enough ugly facts to populate it.

For mathematicians their subject is a science and perhaps the greatest of all. Most scientists, though, use it only as a tool. Like a scalpel, a condenser or a machine that goes 'ping' – or like a chisel or a paintbrush or a lathe – what comes out at the end has more to do with the talents of those who use it than with the equipment itself.

Although astronomy or physical chemistry depends on lots of mathematical machinery, other parts of science – geology, cell biology – manage with almost none (although they do need a lot of 'pings'). Some real geniuses are almost as innumerate as was Darwin (the Nobel Prize-winner Peter Medawar, who worked on the genetics of tissue transplants, when asked what he did when he came across an equation, replied: 'I hum it'). Even Einstein lacked confidence: 'Do not worry about your difficulties in mathematics; I assure you that mine are greater.'

Einstein found the subject hard, and modern students find it harder. It is easy, and may sometimes be justified, to bemoan the fact. But, for the average scientist, how much does it matter? Science is a broad church filled with narrow minds; a refuge for those who cannot do most things but are good at a few. The mathematical illiterates are matched by their many equivalents who cannot extract DNA, recognise a greenfinch or wire a plug. Even Darwin could not do it all; and science has multiplied again and again since his day.

All this is, needless to say, an apologia for my own incompetence. Mathematics is, I was told as a student, like a bikini: it shows what is interesting but conceals what is essential. Like a fool I took the joke seriously. I know the basics, but not much more – and my ignorance has cost me dear. I once spent months in a Balkan tent moving snails around in an attempt to measure natural selection, when a simple statistical test would have told me that to have any chance of success I would need millions rather than thousands of animals.

My experiment foundered on the arithmetical rocks and many others have come to grief in the same place. As a result, many researchers in the life sciences now have a wonderful idea and leave it to someone else (often a computer scientist) to do the real work.

To contract out the hard stuff in this way can pay off, but is not enough. Without some connection to the real world even mathematicians can get it wrong. Lord Kelvin much worried Darwin when he calculated – and he was wrong – that the rate at which the Earth lost heat meant that our planet was young. (Kelvin also made a complete mess of chemistry, with a bizarre taxonomy of the elements based on the geometrical theory of knots.)

As nineteenth-century physicists realised and modern schools tend to forget, science is, sad to say, the mathematisation of knowledge. No longer can we biologists turn to the old joke about the three kinds of people – those who can do maths and those who cannot – but, for the time being, we continue to hum equations in the desperate hope that we are in tune.

The Curse of Oedipus

A quotation familiar to all geneticists: 'It is but sorrow to be wise when wisdom profits not.' Thus Tiresias, the blind seer, when he predicts the terrible fate awaiting Oedipus in Sophocles' play about that unfortunate monarch. More and more, scientists are forced to ask questions much older than science itself. If, as was the case for Oedipus, fate is inborn – in the DNA, to bring the language up to date – what is the point of struggling against it?

Biology, for some people, may turn into Greek tragedy. It exposes truths about which they might prefer to be unaware. Nobody welcomes a DNA-based diagnosis of a death not long deferred, be it from Alzheimer's disease, breast cancer or one of many other inherited conditions that show their effects in middle age. Are we, like Oedipus, learning too much?

Research on Alzheimer's shows how real the danger may be – but also gives some hope, in the distant future, of a treatment. The illness causes a loss of memory combined with random patterns of sleep and waking, rather like those of a baby. Both symptoms place enormous stress on those who care for the patients. They may, it seems, share a common cause – and, just possibly, a common cure.

Sleep and memory are intertwined. Some people lose their memory not because of old age, or Alzheimer's, but because of a stroke or an infection. When taught to play a computer game during the afternoon, they forget all about it within minutes. Wake them up at night, though, and they can describe with great accuracy what they were up to in front of the screen several hours earlier. During sleep the brain cements the day's experiences into the great edifice of recollection that makes each one of us what we are. The process is damaged after a stroke, and demolished by Alzheimer's disease. When the protein formed in the damaged brain cells is injected into rats the animals change their sleep patterns, giving another hint of a tie between sleep and the illness.

The precise timing of the period of rest is also important. In one experiment students learned a computer exercise and were tested on it four days later. One group had three nights of good sleep and one was forced to stay up all the first night – although they could catch up with their slumbers on the other two. In spite of the concession, the second group still did much worse on recalling how the game had worked. Memory, it seems, was consolidated during the crucial first eight hours between the sheets. A later snooze was of no help.

Many genes, differing from family to family, predispose to Alzheimer's (which may well be several illnesses with similar symptoms but different causes). One – the gene for a substance called interleukin-1 – makes a protein present at high levels in the brains of some patients. The gene comes in several flavours, and people who inherit a certain variant have a ten times higher risk of early memory loss than do others.

Deprive a healthy subject of sleep, the great balm of hurt minds, and the level of interleukin-1 goes up. Because the protein's main job is in the immune system (which is itself damaged by sleep deprivation) it also rockets up after infection. In the same way people with flu feel very sleepy, and one of the symptoms of AIDS is intense fatigue – which can, in its later stages, lead to dementia and memory loss.

Sleep, as any airline passenger knows, is under the control of an internal timer set by the sun. It drives daily cycles in the levels of a variety of cellular proteins. They include melatonin and its servant in the body clock, interleukin-1, both of which respond to a change from light to darkness. Alzheimer's patients have abnormal daily rhythms (not helped by the dim light in which many spend their bedridden lives); and the increased agitation and depression shown by some patients in the evening is referred to as the 'sundowner effect'.

The tie between repose and remembrance hints that a single treatment might help in both memory loss and sleep disturbance. Even something as simple as a dose of melatonin, a drug that interferes with interleukin-1 – or that staple of the jet-lagged, a period in bright sunlight – might help those with fading minds.

Such comfort was not available to Oedipus, who, with traditional Greek jollity, pricked out his eyes so as not to see the Sun after he had found the corpse of his mother. She in turn committed suicide when her incestuous relationship with her offspring was revealed. Sophocles, on the other hand, wrote his last play at ninety and choked to death on a grape.

Reflections on India

Remember the Nizam of Hyderabad (not to be confused with the Akond of Swat)? He was one of the world's richest men and the last in a dynasty that stretched back to 1687, when the Moguls destroyed the great castle that once ruled the Kingdom of Golconda. The name was a byword for wealth, for until 1725 all the world's diamonds were mined near by, and the Koh-i-Noor itself, the stone at the centre of the British Crown Jewels, helped pay for the massive fortress that kept the dynasty in power.

I once stood on the summit of that huge ruin, its walls seven miles round, and surveyed what Hyderabad has become – a city bigger than London, its polluted streets jammed with beggars, traffic, cattle and crocodiles of well-dressed children on their way to school.

All tourists in India suffer a pushy guide with his dubious tales of secret tunnels and the like. Even so, my own escort's stories rang a bell. Just inside the entrance to Golconda, beyond its gigantic spiked portal (the spikes kept off the elephants) is a gatehouse with an intricately modelled roof and a narrow slit for an exit. To clap hands at a precise spot in the centre gives a long series of staccato echoes, audible at just that point.

An hour later, at the summit, half a mile on and several hundred feet up, a wave by my guide to his mate in the gatehouse prompted the sound of a sudden clap, which could be heard only where the guard of the inner sanctum once stood. In the old days the signal meant trouble and was a hasty call to arms.

'First telephone!' said the guide, and he was more right than he thought. The echoing gatehouse with its well-placed slit demonstrates what physicists call total internal reflection. Light does the same thing. As a ray passes from air to water or vice versa its direction changes, but as the angle becomes more acute it no longer refracts into the new medium but is reflected back from its surface. Diamonds – with a refractive index much greater than that of water – depend on the same process: a cut stone gathers in light from its whole exterior and after a series of internal reflections flashes fire from its face. The slit in Golconda's gatehouse matches the facets on the stone for sound rather than light. The city's jewellers were famous for their skill, and – not inconceivably – those who cut Golconda's diamonds designed the castle's alarm system on the same principle.

Hyderabad is sometimes called Cyberabad, for the city is the focal point of India's high-tech industries and of the call centres that service many British companies. The economics of telephony has been revolutionised by fibre optics, a system that transforms sound to light and passes it down thin strands of glass. So narrow are the fibres that the beam shows total internal reflection even as the pipe carrying it goes round corners, and the signal passes with almost no loss down a cable that may be many miles long.

Biology has taken up the idea. The total internal reflection fluorescent microscope lives up to its name. A laser beam shone in from the side of the specimen strikes a living cell at an angle sharp enough to cause internal reflection. By labelling the cell's molecules with a fluorescent dye that glows when struck by the laser the microscope can view an exquisitely thin section. It tracks the movement of single viruses as they invade, of muscle proteins as they contract and of the cell membrane as it controls the flow of hormones in and out.

Hyderabad is a world centre for the study of those processes and of much more, from DNA fingerprints to the structure of proteins. Its remarkable and up-to-date Centre for Cell and Molecular Biology houses much of the technology of modern genetics on its extensive campus and has a dedicated staff to match. India may be short of natural resources, but it saw the importance of brainpower long ago. I spoke to many students and young scientists at CCMB: well educated, hard at work and keen to make a career in science, which, in their native land, is a relatively well-paid and well-structured profession. Already the country is at the forefront of stem-cell research and is pushing towards leadership in bioinformatics, the interface of computer science with genetics.

In legend the streets of Golconda were paved with gems. Although the country's own seams have been exhausted, the Indian diamond industry now processes nine out of every ten stones mined in the world in what has become its second-largest export earner. Judging from the Hyderabad centre, biotech may be next. The thought might lead to a certain internal reflection by scientific planners on this side of the globe.

Cracking the Code

In 1891, in the village of Rennes-le-Château, on the French side of the Pyrenees, the local priest, the Abbé Saunière, uncovered a coded parchment. He deciphered it and found – according to which source one chooses to believe – the treasures of Jerusalem, the Holy Grail or Jesus' tomb. At once the Abbé gained mysterious riches, rebuilt his church and graced it with a most devilish statue. Rennes today is filled with New Agers, led by the cipher to an endless search for buried treasure. Some claim to have found many other such messages hidden in the Bible, ready for those who find the key. The internet also contains hundreds of prophecies, from the collapse of the Twin Towers to the defeat of Saddam Hussein.

Cryptography has a long history in biology, too; and can give results just as spurious of those of Rennes. The DNA cipher, the famous three-letter code, was cracked by Francis Crick and others in a classic experiment. Before he succeeded at the bench, Crick tried some molecular cabbalism which showed how life's code ought to work. He knew that there were four DNA bases – the letters A, G, C and T, for short – that code for twenty amino acids (which have themselves been given letters – H, I, D, L and so on – to help computer searches of the database).

The idea of a cipher set out in threes made more sense than one in twos, which would give no more than sixteen combinations of four letters. A bit of thought gave Crick an answer so elegant that, he thought, it must be true. He suggested that only a few of the sixty-four three-letter combinations make sense. The code must, he deduced, be arranged in such a way that any shuffling among the valid triplets is meaningless. To put his case into familiar words, if CAT and DOG make sense, then CTA, TCA, GDO, DGO, ATD, TDO, OGC and GCA – some of the other ways in which those letters can be ordered (reversals don't count) – do not.

To apply this law to all possible combinations of the four bases showed – surprise, surprise – that the largest number of valid triplets is twenty; just the same as the number of amino acids. That must, Crick thought, be how the genetic message works. His ingenious logic had revealed the secrets encrypted into the DNA and soon, he was sure, it would lead to the deepest mysteries of life.

That beautiful theory faces, alas, just one problem: it is entirely false. His theory has been described as the cleverest idea in the history of science to have turned out to be wrong. The code of the genes is in fact much less elegant than the one deduced by Crick. The message is indeed written in triplets: each amino acid encoded by three letters of the DNA bases. However, all sixty-four possible combinations are used, because some amino acids can be spelled in different ways, with more than one word for the same thing, and because some triplets are used as punctuation marks.

The genetic code, untidy as it may be, is universal and descends from the origin of life. The Book of Genesis – another version of how life arose – itself contains, some say, another cipher. Its divine cryptography has been cracked by a group of Israeli computer experts. They looked at the spaces between letters in this first chapter of the Bible. If one starts with, say, C, the third letter in the alphabet, and reads every three letters thereafter, great strings of characters emerge. Now and again, by reading up, down or crosswise, they spell out names. Some are of rabbis born long after the book was written – and their names are close to dates that appear to predict their births. This simple result sold a million copies of a book on Bible codes.

Cynics have tested that approach on other great books. Again the results probe the limits of the credible, for they prove that *War and Peace* has rabbis skulking in the text, that *Moby Dick* predicts the death of Princess Diana and that a certain Bill Gates is tucked away in the Book of Revelation. Needless to say, none of this is a message from a higher source. The results emerge because so many tests for a fit were made and because the rules used to recognise them were even laxer than those of the genetic code. The Israeli experts and their Bible code were far too generous in allowing for variant spellings. As a result, their millions of computerised attempts to find a match for names could do nothing but succeed – as they do for any book. Bible codes, like Francis Crick's Bad Idea, arise from mere arithmetic optimism.

Even so, by matching the code in the millions of proteins now sequenced with the *Complete Oxford English Dictionary*, cryptographers using the letters of the amino-acid code have found the word 'HIDALGISM' – the art of being a *hidalgo*, a Spanish nobleman – spelled out in certain proteins, and Rennes is just a few miles from the frontier! There lies, without doubt, the key to Abbé Saunière's treasure.

Engineering the Arts–Science Divide

The central atrium of Tate Modern specialises in Big Things. In the year of the millennium it featured a massive metal construction covered in wheels, levers and mirrors with a strong resemblance to a giant experiment in particle physics. It was not engineering, but art: a piece by the American sculptor Louise Bourgeois. It allows, said the catalogue, those who experience it to undergo 'significant conversations and human confrontations'.

The aesthetic ziggurat left me cold. A spectator in the same spot twenty years earlier would have had his attention riveted by a more formidable object: the 120-megawatt steam turbine at the heart of Bankside Power Station. When at work, such machines emit a screaming roar (and I lost some of my hearing in the year I spent as an acolyte of a smaller version) and now and again, like all turbines (but not many artworks), burst into random, dangerous but inspirational flames.

Under maintenance, its delicate and expensive guts revealed, a steam turbine is a thing of beauty; and I had many significant conversations and confrontations as I passed tools or tea to those at work to keep the beast alive (it was, after all, the 1960s).

Many artists have tried to fuse the passionate subjectivity of art with the cold neutrality of science. Seurat, for example, employed the theory of colour to develop a new way of painting based on dots and many artists were expert anatomists. The dialogue between the two talents was productive indeed. Now, though, science has been hijacked by artists in a most eccentric way; not as essence but as embellishment. To use science as an aesthetic device to add flavour to objects that have otherwise nothing to do with it is an excuse for a deliberate failure to understand; for a satisfaction with the superficial and a refusal to look even just below the surface.

New York's Whitney Museum Biennial of 1997 was filled with such references. Biology got a look-in with *Ignis Fatuus*: at first sight an explosion in an abattoir, but in fact hundreds of party balloons knotted together to resemble the guts of a monstrous cow. Near by a foot-wide moth settled on a white wall; a match in plastic, brass and exquisite detail of the real thing. Science was used to other unexpected ends. A series of oils by Matthew Ritchie was grouped under the title 10^{-38}, its canvases adorned, for no obvious reason, with terms like 'scalar fields' and 'limbic system' and with fragments of the periodic table of the elements. One piece showed, without explanation, figures based on lampbrush chromosomes: giant pieces of DNA found in frogs. How many of those who admire the work recognise them? And why was a picture of pylons given the mathematical title *Once Inverted, now Exponential*?

On this side of the Atlantic the Royal College of Art has also tried various arts–science events. Its successes and failures had lessons for both fields. In one exhibition the mathematician Roger Penrose showed several drawings, from non-repeating patterns of tiles to one of Galileo as he threw large and small rocks off the Leaning Tower of Pisa. Some were based on the Mandelbrot set; fractal images interpreted as art. In a neat complement, the sculptor Antony Gormley used such geometry (with the help of three miles of steel) in his *Quantum Cloud*, which hints, from certain angles, at the presence of a human form inside what appears at first to be a random cluster of bars. Science, it seems, still can offer something to the aesthetic imagination.

Penrose has a penchant for long titles: *How the Riemann sphere translates the abstract complex number-weighting factors ('amplitudes') w, z, into a geometrical direction. For a particle of spin 1/2, this becomes the direction of the particle's spin axis* – which sounds like the Whitney's *Once Inverted, now Exponential* but, unlike that, makes sense (to mathematicians at least).

According to Theophrastus, 'Art imitates Nature'. The phrase is familiar enough, but the rest of the quote tells us: '. . . and sometimes produces very peculiar things'. They often emerge from today's interactions between science and art. The successes come not from forced and artificial syntheses but when the two keep a respectful distance. The worst failures are in the lazy attempts by artists to parody science with no thought as to what it means, or by scientists to persuade themselves of an aesthetic talent that they lack. Art does not need science, although it can use it; and science may produce beauty almost by accident. Both parties should admit that to travel in a foreign land without speaking the language is, too often, to make a fool of oneself.

A Postilion Struck by Lightning

One of an author's secondary pleasures is to be sent the translations of his own works. To see a book with your name on it even when its contents are impenetrable (as in the Hungarian edition) gives a definite frisson. In many languages it is possible to identify one's own words, reborn into another world.

It is much harder to go the other way: to do the translation from English itself. Languages may be related, but each has a personality of its own. The German versions of my books are a fifth longer than the original, which makes them even more ponderous than they already are, while the Japanese editions are elegant, slim and quite impossible to understand.

New tools on the internet make the language shift at the touch of a button. Once they just lined up words in each tongue, which gave a garbled sentence at best. Now they incorporate syntax; look for subject–verb–object in the original and use the target's grammar to generate the new sentence. They have much improved, but are still far from perfect (a sentence which, when machine-translated into French and back again, reads: 'They improved much, but are still far from the perfection').

Such an approach works because the electronic brain has been made to compare huge chunks of material with a version done by a linguist. The programs involved are pretty technical, with disambiguation, lemmatisation and such great names as these, but the idea is simple: plod on and build a vast vocabulary until you crash through the language

barrier with brute force. The famous (and no doubt apoc-
ryphal) translation of 'out of sight, out of mind' as 'blind
idiot' is cured by telling the computer to look out for the
ambiguous phrase and others like it, and beware.

The process is slow and expensive and for less common
languages makes no economic sense. As a result, my translated sentence put by the machine into Hungarian (whose
computer database is far smaller than that of French) now
reads, bafflingly: 'They have many improved, but there is
silent wide that is perfect', which is beta-minus at best.

The machines are taught, but they do not really learn.
Infants, in contrast, pick up the local lingo fast, and not
word by word but almost sentence by sentence. As readers
of this volume have no doubt noticed, my own native
tongue is not English. I made the shift from Welsh at around
the age of six. My memories of doing so bring to mind the
Winnie the Pooh books and suddenly being able to fit the
text to the pictures, with no conscious effort to learn one
language by using the rules of the other (which is fortunate,
as Welsh grammar is bizarre). I do remember one problem
which hints at hidden linguistic depths, for Welsh has a
word – 'glas' – whose meaning hovers between 'blue' and
'green', and for a long time I was puzzled by the distinction
between the English words for the colour of the nation's sky
and of its grass.

In translation, brute force is giving way to an attempt to
persuade machines to use their own efforts to learn a foreign
tongue, with much less help from a bilingual minder. The
new approach entails feeding in modest-sized blocks of
English text and their translation, together with some clever
statistical instructions based on combinations of words. The
computer picks up the linguistic rules as it goes along rather

than being told every single detail of the new grammar by its owner. In time it builds up longer and longer groups and identifies those that tend to occur together – blue sky, green grass, rather than the other way around – to help in its task. It can see words in context ('still far from' turns up in English more often than 'silent wide that is') rather than just fitting them on to a rigid framework. Soon it begins to manage on its own. The method, say its proponents, will crack a hundred languages – Uzbek, Tagalog, Ossetian and more – within the next five years.

I have almost forgotten my own native tongue – and the process of loss was almost the inverse of learning a new language. Vocabulary went first, then grammar, until all that was left was the accent (which is the hardest thing to perfect when an adult learns a foreign tongue). I can speak utter nonsense in Welsh, but perfectly pronounced.

What I need is one of those much-promised gadgets to take in speech in one language and blare it out in another. However, as all Welshmen know, their native tongue is only spoken when English people are in the room, which removes the need for a Celtic version.

Oysters for Lunch

For breakfast a six-inch slice of cheesecake. Lunch: deep-fried oysters with mayonnaise, followed by a hearty dinner of stuffed veal. (I was forced to turn down the toffee ice-cream dessert as I could not manage it.) That is not at all my usual diet, a glum but healthy mix of muesli, salad and salmon, but the input of the burghers of Milwaukee, where I spoke at a fund-raising meeting at its Medical College. The talk was over breakfast at 7.30 a.m. and I got a mild titter by describing metabolism as a mechanism for turning cheesecake into hot air.

Almost all the participants at the nerve-wrenchingly early event were trim and fit, but a stroll through the streets showed them to be the exception. For the first time in history the rich are thin and the poor fat. Everyone knows that a wave of obesity is sweeping the globe but I was shocked to see the tsunami of lard that has enveloped the Land of the Free.

The average US citizen uses as much energy each day as does a blue whale, and even the Brits are well above porpoise level. Every twenty-four hours another thousand tons of fat settles around the American stomach, adding a hundred billion dollars each year to the health budget. Worldwide, a quarter of a billion people are overweight, with the worst yet to come as the new cohort of flabby children grows up. They may be the first American generation to have shorter lives than their parents.

One danger lies in diabetes. Not the childhood kind which can be controlled with insulin but the adult-onset form, which leads to heart disease, blindness and even amputation. The problem comes from the body's inability to respond to (rather than to make) the hormone and is far more common among the plump than the slim. Unless lifestyles change, one American in three of the cohort born at the turn of the millennium will develop the illness during their life – and the figures for Europe are not much better.

In the American way, the nation is trying to tackle the crisis by throwing money at it. The Medical College of Wisconsin uses rats (certain strains of which are very susceptible to diabetes and its associated ailments) to track down the genes responsible for such conditions and – with luck – discover their equivalents in the human genome. The research is just part of the huge effort devoted to the genetics of common diseases. Admirable though it may be, what actually is the point? It might seem obvious. Once people fell to external enemies – infection, violence or cold. Now we succumb to the enemy within – our inborn imperfections. In the modern world almost all of us die from a genetic disease, for cancer and heart disease have, like diabetes, a considerable inherited basis.

Surely, then, the answers lie in the DNA, which, even if it does not come up with a cure, will at least warn those most at risk to change their ways. Life, alas, is not so simple. First, such illnesses are complicated. Adult-onset diabetes can emerge from problems in the pancreas (the site of insulin secretion), the liver, the brain and elsewhere. Dozens of genes may be involved for each illness (already more than a hundred for diabetes, with the rats poised to reveal more) and because they vary from population to population, or family to family, there is no universal test. Even those told of the risk tend to take little notice. And when it comes to prediction (a question which exercises the health insurance industry) for most people the role of genetics is trivial compared with that of diet.

As was true in the battle against smoking, the answer to obesity lies not in the flesh but in the soul. The American Society of Bariatric Physicians (they study obesity rather than being obese themselves) has proclaimed a fatwa against fat. Many things weigh on the nation's mind. Some crusaders are giving out pedometers to ask people how many paces they take a day. Their Ten Thousand Step Program aims to promote moderate exercise and has had some success. New estates have gained planning permission by separating people from cars to force their inhabitants to walk farther and get thinner. Restaurant menus warn of the dangers of certain dishes and the Campaign for Breakfast (and half of all Americans skip that meal) publicises the fact that a full stomach in the morning reduces the urge for a snack later. And no, I won't have another slice of cheesecake, thank you.

The Joy of Sludge

Water is remarkable stuff. It acts as a shock absorber in the Earth's rocky relationship with the Sun. Solar energy streams in, the amount varying with distance from the equator, with the seasons and with day and night. Water, with great efficiency and without complaint, distributes it around the planet in the form of clouds, rivers and ocean currents. As it does the job it damps out what would otherwise be fierce local, annual and daily swings in temperature.

Before man got into the act, most water worked on the local scale. In a woodland, for example, each tree acts as a pump which sprays vapour into the atmosphere during the day. That in turn condenses on the vegetation at night, cools the air and stabilises the temperature (which is why forests have less variable climates than do open fields). In any mature system, such as a forest or a marsh, or even a rice-paddy, groundwater is remarkably static; a little blows away as moisture or flows off in streams, but most is recycled within a few hundred metres of where it started. As a result, not much of the landscape is washed away.

In the modern world a new and liquid market has emerged and water travels much farther than it did. The softly running Thames is brown not because it is full of nameless wastes as once it was, but of mud. Glorious the river may be; but its swirling silt represents a constant slow withdrawal from England's soil bank. The landscape is being washed away, and the balance in the geological deposit account is sinking fast. A modern English farm can lose half a ton of minerals from each acre each year from its ploughed fields. The Romans, when they built London Wall around their settlement, which was sited in a forested landscape, paddled in a stream far clearer than the modern Thames. Even fifty years ago the rate of soil loss from southern England was much less than it is today.

The problem is universal. Fields are not closed systems; they have to be drained or irrigated as the seasons change and often become sodden or parched. Water levels shoot up and down as millions of gallons are pumped in or stream away. All this has dire effects on the soil, which breaks down at a far greater rate than before – and is swept off for miles in the newly empowered rivers.

Lake Trummen in Sweden has had its bottom much examined in the search for the history of silt. At the end of the Ice Age, with a bare and shattered landscape round about, it gained a millimetre of sediment a year. Within a few thousand years the figure dropped by nine-tenths as forests stabilised the land. The nineteenth century saw a dramatic change. The amount laid down shot up by a hundred times as man, with no thought for the future, stripped the trees for farmland. In the 1960s the lake almost died. The muck was pumped out and spread on the local fields – which rescued Lake Trummen but would be far too expensive to try in most other places.

The French face a similar problem. The island of Mont St Michel has been home to a monastery for more than a thousand years. Now large parts of the sandy tidal estuary have turned into a green field with a ditch in it and the island has become a peninsula, thanks to the mud brought down by the local rivers. The state, ever anxious about its patrimony, has an ambitious scheme to save the place and plans to build a dam opened at each tide to give a great surge to flush the mud out to sea.

That scheme treats symptoms rather than causes. The only real hope for Mont St Michel – and for many other parts of the European landscape – is to restore the land's own ability to act as a sponge. France has had great floods in the past few years, but they are as nothing when compared with what happened a century ago. In the 1850s the Rhône was often blocked by the debris that came down on its furious torrent and floods devastated many of the towns on its banks. So much of the mountain forest of the Cévennes and the Alps had been destroyed that large areas were referred to as 'Le Désert'. In the face of bitter opposition from farmers, millions of trees were planted and, a century later, the problem was to a large degree solved. As to whether the wheat barons of England can be forced to give up their subsidies in the interests of the cities they claim to despise, no doubt we must wait until the next great inundation for an answer.

A Change in Political Chemistry

Chemistry: the word that turns most people off science for life. I remember the grind of A-level, with mindless learning of formulae in the blind hope that they would be in the exam and the boredom as one committed to memory the best way to extract silver, synthesise ammonia and crack petroleum.

I abandoned the subject with immense relief but, thirty years on, went to a chemistry lecture. It had an audience of five hundred schoolkids, who (unlike my contemporaries) laughed, clapped and asked for more. The lecture dealt with polymers and showed how strings of molecules can bounce, bubble, flash in the dark and even flow uphill.

The place made the occasion. I was at the SASOL Science Festival, in Grahamstown, South Africa; a week-long event with tens of thousands of visitors aged from six upwards, from townships (under a scheme to help disadvantaged pupils sponsored by Mercedes-Benz of South Africa) to expensive private schools in the suburbs. Some students had been driven hundreds of miles overnight for a chance to see science in action.

It was in many ways an inspiring event, with speakers of international standard. From an attempt at a medieval siege catapult to alarming displays of the dry-ice-powered bottle-rocket, long banned from British schools, and from bat walks to teachers' workshops, science unfolded itself to young people who had never seen an experiment before.

SASOL, once the South African Coal Oil and Gas Corporation, was built on chemistry mixed with conflict.

Apartheid South Africa had no oil but plenty of coal and the company's scientists managed to develop new ways of changing one polymer into another, coal into oil. The country is still at the frontiers of chemical research, but is keen to get more of its people involved in science than was possible in its sterile days of linear thinking, when racial separation outweighed what the complicated modern world demands.

South Africa proves that history does not travel in straight lines. A small political change may have gigantic consequences. Mathematics shows why. Crowds are a perfect example. They stay calm for most of the time, but can feed on themselves and burst into unexpected violence. Chemistry, too, is in a new era in which reactions are seen not as linear systems but as complicated and interacting universes that adapt to the world they create. The laws of life or physics may be simple, but the world they make is convoluted indeed, with many swerves on the way, in the form of both revolutions and chemical plant explosions. Molecules can behave like people in crowds. For both the buzzword is 'complexity': a system in which small changes have great and unpredictable outcomes. A chemical reaction, a mob or a nation is poised to take one of many pathways and which one it follows depends only on the initial shove.

Chemists gained their first insights into the new chemistry by turning for advice to those who make computer models of the brain. Artificial neural networks – programs trained to recognise patterns – were developed to help understand its almost impossible intricacy, but are now used to study chemicals that might be used as drugs, to develop new compounds from as yet unthought-of elements and even to make new polymers that, like the wood turned first to coal and then to oil, are better than anything made in a factory. Such computer models even help us to understand riots.

The SciFest talks were not just on chemistry, but on the brain, water and the *Titanic*. As the meeting went on, a seismic shift took place in the audience; although the morning's public exhibits were filled with black children, those in the talks, later in the day, were mainly white. The reason was obvious, for it cost a pound to get into a lecture and most black schools could not afford it. Just as in a chemical reactor, a small shift in starting conditions had a big effect on the end product.

And that is almost a definition of politics, in South Africa more than most places. I talked to a teacher from Port Elizabeth who had just been shocked to discover that her new township school, built using a set of prefabricated structures, had been stolen – the whole lot, down to the foundations – by an armed gang who had come with lorries and driven it off. In spite of all the optimism on view in Grahamstown, her loss was a reminder of how complex is the chemistry of a society in transition, and how chaos is never far away.

The Sage of Lichfield

The late Auberon Waugh's Bad Sex Awards go to the most cringing and awful description of the generative act in a modern novel. Every prize-winner should be burned on the grounds that it might deprave and corrupt the literary style of those who read it. Two hundred years ago these lines would have been in the running:

> *Each wanton beauty, tricked in all her grace,*
> *Shakes the bright dew-drops from her blushing face;*
> *In gay undress displays her rival charms,*
> *And calls her wandering lovers to her arms.*

Literature indeed, compared with the dross written today; but belittled in its own time for 'Unsuitableness of subject, rhetorical extravagance and convention of phrase'.

Fair criticism, perhaps, when we compare it with the authentic botany of Parnassus:

> *No, no, go not to Lethe, neither twist*
> *Wolf's-bane tight-rooted for its poisonous wine;*
> *Nor suffer thy pale forehead to be kiss'd*
> *By nightshade, ruby grape of Proserpine . . .*

Keats, in the *Ode to Melancholy*; but what of those other iambic pentameters? They were written by a medical man who invented the photocopier, the speaking machine and rack-and-pinion steering, and was, for a time, seen as England's greatest poet, with an influence that reached to

Keats himself. What is more, his grandson appears on a British banknote.

Erasmus Darwin was England's Leonardo da Vinci, with interests that ranged from physics to botany. *The Loves of the Plants*, published in 1789, was a work of science in 238 pages of verse. It deals with the 'sexual system', a method of classifying plants based on the balance between male and female parts and illustrated with Arcadian images. The 'wanton beauty' of his protagonist represents species in which males and females are separate – which means that pollen (the wandering lovers of the poem) must travel to meet egg.

No wonder the Victorians saw his lines as unsuitable; but Erasmus, like all great scientists, was far ahead of his time. His poem notes the 'gay undress' of the females and their competition for mates by displaying their 'rival charms'. That is almost right, but as his grandson Charles pointed out, males rather than females suffer most in the struggle for sexual success. As has often been said, from a plant's point of view a bee is a flying penis, needed to transfer pollen to egg. In the many plants that are both male and female ('Sweet blooms *Genista* in the myrtle shade, / and ten fond brothers woo the haughty maid') a single visit fertilises the egg, but male success depends on pumping out as much pollen as possible. As a result, most hermaphrodite flowers, feminine though their public image might be, are (like the mandrill's bottom) silent screams of masculine passion.

Sexual selection, as Charles called it (and Erasmus himself was no mean performer, with two wives, a mistress and twelve children), can drive the origin of species itself. Creatures with showy males, from flowers to ducks, tend to be split into lots of different species, while those with more modest swains are much less subdivided. As another hint of the importance of sex to evolution, the fossil record reveals an explosion of change when the first flowers appeared and plants joined the sexual battle for attention. In the monkey flower a single mutation that changes flower colour from pink (which is attractive to bees) to red (favoured by hummingbirds) has started a new species almost in one step. Sex (much as Auberon Waugh would have disapproved) really does make the world go round.

Both Erasmus Darwin and Keats would have been delighted to learn as much. The *Ode to Melancholy* was written in Hampstead, in a house that, in 2000, celebrated its seventy-fifth anniversary as a museum. Thanks to English Heritage, the building is under restoration and is a site of pilgrimage for poetry lovers. Erasmus had to wait longer for a memorial and his home opened to the public only in 1999. It stands near the cathedral in Lichfield. A solid Georgian structure built around an older core, it has, with the help of the National Lottery, been wonderfully restored. On show, among other memorabilia, is Britain's oldest bottle of beer, brewed to celebrate the great man's christening.

The home of poetry is secured by those hard-headed but clearly poetic men of the Corporation of the City of London. The future of the Erasmus Darwin Centre is less certain, for it depends on private donations, most of them small. If the manufacturer of photocopiers, of cars, of hi-fi (or even of pornographic novels) made a tiny gesture to their great progenitor the memorial to the lesser poet but great scientist could, one day, be as secure as that to Keats himself.

A Brief Constitutional

The Declaration of Independence formed the modern world. Noble though its language may be, the single sheet of parchment is ambiguous indeed, for it appeals to 'the laws of Nature and of Nature's God', to truths defined to be 'self-evident' and to many other rhetorical flourishes familiar to lawyers arguing a dubious case.

The Founding Fathers used legal language to force their views upon numbers of loyal North Americans of British descent who would rather have stuck with the old system – but so what? It worked, and the seven Articles of the US Constitution are a model of clarity.

The Treaty for a Constitution for Europe, in contrast, runs to 265 pages. It begins with vaporous guff (Europe is 'the great venture which makes of it a special area of human hope') and then surrenders to the lawyers. Science gets a look-in ('The Union . . . shall promote scientific and technological advance'), but anti-science is allowed a subtle, but potentially fatal, crack of the whip.

Just after Abolition of the Death Penalty but before the Prohibition of Torture and of Slavery, the draft Charter of Fundamental Rights insists on 'the prohibition of eugenic practices'. This sounds self-evident – noble even – but what do they mean?

The word 'eugenics' – from the Greek for 'well born' – was invented by Francis Galton, Darwin's cousin, to reflect concern about a supposed decline in the quality of the human race. He favoured fertility for fine minds and sterility for the stupid; and his ideas led in a direct line to the horrors of the Nazi era. Thanks to a speech by Josiah Wedgwood MP a 1918 attempt to institute sterilisation in Britain failed, but there was plenty in the United States, and Scandinavia was not immune.

The slogan 'Down with Eugenics and Up with the Constitution' seems unambiguous, but is not. I have a special interest in the issue, for I am, for my sins, president of what was once the Eugenics Education Society (a century old and reborn as the Galton Institute) and it would be nice to know why I am to be abolished. I did write to Brussels for clarification, but answer came there none. The Institute is nowadays concerned with social aspects of genetics; how people respond to advice; and how reproductive technology can improve lives. The e-word, though, is, in Europe's eyes, enough to damn us.

Why the fuss? Large parts are based on prejudice incited by the Catholic Church. Some of the toxic mix involves simple ignorance about what science can do. The press is full of phrases like 'designer babies' and 'playing God' and one target of the Euro-edict is pre-implantation genetic diagnosis. This allows parents at risk of a child with inherited disease to choose an embryo shown to be free of the gene. Their baby, no doubt, is 'well born'. Galton would approve – but will the Constitution? On a strict construction, no; and the clause is an attempt by a religious minority to force its views on the secular European majority.

In-vitro fertilisation is also in their sights. There are great national differences in attitude. The German Basic Law – 'Human dignity shall be inviolable' – applies, they maintain, to the egg, which much limits what clinics can do. The British are more tolerant, but the NHS is mean and many turn to the private sector (which costs more in the end as such clinics tend to implant many embryos, causing multiple births, health problems among children and expense for the taxpayer). The French are pragmatic and the state pays for treatment – but that nation is cautious about the creation of embryos for research alone.

The Constitution makes another avowal: 'Scientific research shall be free of constraint.' That might tempt scientists to turn to the Union for a grant – but some will be unlucky, for elsewhere in its rambling pages it says that 'creating human embryos solely for the purpose of research or stem cell procurement is forbidden'. The practice is legal in Britain, but will not be funded from Europe, which sounds pretty much like a constraint to me.

Although to have laboratories policed by prayerful purists would be a disaster, there are real reasons for concern. We are not keeping close enough tabs on the children born by IVF. Might there be long-term effects? Mice produced in this way are more confident in their behaviour than are others, but have shorter memories. With more than a million test-tube babies now in the world there have been few attempts to follow them up.

For real dishonesty one looks to the colonies. The United States has made a complex patchwork of contradictory state and federal laws on stem cells. It has moved on from that shambles with the Unborn Victims of Violence Act, which provides separate protection to a foetus should its pregnant mother suffer a crime. Its real agenda is an attack on abortion, for to give legal status to the unborn undermines a woman's right to control her own body. Any attempt to use the European document to introduce the anti-eugenic notion to America would curb civil freedoms that have taken years to establish.

Of course, those who drafted Article II-3, sub-section 2b never allowed such thoughts to cross their minds, but perhaps the people of our damp and liberal island should think twice about their new Constitution and thank the French and Dutch for killing it off for reasons that had nothing to do with science.

Requiem for a Mathematician

My old friend and colleague Cedric Smith died not long ago at the age of eighty-four. He was a member of that disappearing breed of academics who could immerse themselves in work, rather than in administration. Like many mathematicians, he was fond of jokes. Thus, what did Jesus mean when he said: 'Heaven equals $ax2 + bx + c$'? I, as an innumerate biologist, had to be let into the secret; it's a parabola.

Cedric Smith invented a new system of arithmetic (which worked well with a bit of practice) based on counting from one to five, and replacing all numbers greater than five with the same number subtracted from ten and printed upside down. He also developed the standard statistical methods used to map genes on to chromosomes and wrote a solid text called *Biomathematics*. His most famous results appeared not as the works of Smith, but under the pseudonym of one Blanche Descartes, a lady still much referred to in the mathematical literature.

The mythical lady was in fact a committee. She consisted of a non-empty set (as mathematicians would say) drawn from the four members, Cedric included, of Trinity College Mathematical Society, who met when they went up to Cambridge in the 1930s. All became famous, and one among them, Bill Tutte, was an important part of the Bletchley Park Enigma Code decryption team.

Blanche (or her avatars) published on a great variety of subjects. One afternoon the group was interested in the problem of whether it was possible to cut a square into a number of smaller squares, with no two of them the same.

After a lot of scribbling it turned out that it was, and the squared square had been invented. The notion seems pretty abstruse, but squared squares are of great value in the design of electrical networks. The smallest possible version – not discovered until 1978 – consists of twenty-one separate squares, each one different, which can be assembled into a single unit.

The group also came up with 'Blanche's Dissection', the simplest way to divide a square into rectangles of the same area but different shape. Dozens have been found and such tiling problems, as they are called, have become an important theme in modern mathematics. In what must be more than a coincidence, Roger Penrose, the son of Lionel Penrose, one of Cedric's close colleagues of the 1950s, became a professor of mathematics at Oxford and invented a system in which an infinitely large flat surface can be covered with a pattern of tiles that never repeats itself. Sixty years after the Dissection of Squares, Cedric (or Blanche) moved on to the same trick with triangles.

Another of their interests involved what became known as the Counterfeit Coin Problem. You are given a shilling's worth of pennies (twelve, for younger readers), one of which is counterfeit and differs from the others only in being the wrong weight. Using a pair of balance scales, what is the smallest number of weighings needed to pick out the false coin? The answer is three; and Blanche came up with a solution that works for any number of coins. Not satisfied with that, she generated a mathematical series that made infinite numbers of fairly tuneful tunes; a trick of the useless kind that so much delights true mathematicians.

The counterfeit coin game may also seem a little obscure, but was one of the first developments in search theory, a branch of mathematics now much used in computer science, economics and of course in looking for crashed aircraft and lost mountaineers.

The last pleased Cedric, for he was a Quaker and passionate pacifist, who was much involved in the Peace Studies movement and in the mathematical analysis of wars and what might be done to stop them. He was much less happy about the other applications of search theory, which are military and were no doubt used in the attacks upon Afghanistan, Iraq and the other places that dare to disagree with the latest disciples of Jesus himself.

In the final paragraph of *Biomathematics* Cedric Smith wrote what can be read as his epitaph: 'Mathematics needs to be wisely guided. Like other instruments, it can be used properly and with discrimination, or foolishly and inappropriately. Even the finest and most effective instrument can be put to base as well as to noble purposes, to impoverishment and destruction as well as to the enlargement of life and the creation of beauty.'

Blanche Descartes has continued to publish papers (albeit under the names of her individual members) until the present day; and although Cedric promised to reveal one day how she got her own name he died, alas, before Blanche's secret could be revealed.

Sam Weller's Christmas Lunch

December is the unhealthiest month, with more burns, more asthma and more suicide than at any other time of year. Most of the problems are due to the Dickensian flummery of combustible trees, smoky fires and seasonal despair.

In France Christmas has a different flavour. Many families settle down to a festive meal not of bland bird but of oysters. And that was once true in England, too. Oysters were cheap, delicious and universal. As Sam Weller says in *Pickwick Papers*: 'It is a wery remarkable circumstance . . . that poverty and oysters always seem to go together. Blessed if I don't think that ven a man's wery poor, he rushes out of his lodgings and eats oysters in reg'lar desperation.' In those days only the rich could afford turkeys (or geese, which came to the same thing).

Today decent English oysters cost more than a pound each and in protein terms are fifty times more expensive than Bernard Matthews' finest. But why? How did a dish once as humble as cod and chips become the preserve of the affluent? The decline of the British oyster is a tale of economics, of European policy and of ecological mystery.

In 1850 half a billion oysters passed through London. Thirty years later the number had fallen by nine-tenths. Alarm at the collapse led to the setting up of a Royal Commission. Its advice was simple and seemed to be scientific. There was no point in restrictions on fishing. As each oyster could produce eight hundred thousand eggs each time it spawned, the crash was not due to man's activities. Biological authority had spoken, and authority it was, for the prime mover was Thomas Henry Huxley, Darwin's great defender. Let the market reign: nature's balance would put things right.

In France the state stepped in. The shore belonged to the government and was rented out under strict conditions. Fishing was restricted, sometimes to a few hours a year. Let *les rosbifs* try laissez-faire; in France, with echoes of today's beef, the oyster – '*aliment complet, médicament précieux*' – was under the protection of the Republic; and English science could go hang.

The rest is history. Nowadays the French produce two billion of the creatures each year (two hundred times more than us), three-quarters of them sold at Christmas. We still do not understand what limits the animals' numbers in nature, but overfishing certainly has a drastic effect when humans move in. As so often in ecology, the obvious is wrong and nature's interactions prove impossible to predict.

Today's biologists are often reminded of that lesson. Marion Island in the Antarctic was once much visited by sealers (who ceased their visits only when they had killed off their own prey). A few mice escaped from a sealing ship and, once ashore, turned for food to a flightless moth. In 1948, in an attempt to control them, the South African government introduced five cats. These ran wild and did such good work on the rodents that they were forced to turn to the island's fourteen species of seabird. Soon cats were killing half a million birds a year. So many died that less guano was produced than before and lush grass was reduced to bare ground.

In 1977 a virus was let loose and all the cats died. Some birds returned, but others did not, for the mice, again abundant, had eaten so many insects that many birds starved. In the latest turn of the ecological screw the vegetation is under threat from debris as the insects once responsible for recycling it have been driven to extinction. Ecologists plan to wipe out the mice for good but what will then happen is impossible to say.

T. H. Huxley was as confident about cod as about oysters. In an 1866 report he wrote that: 'Any tendency to over-fishing will meet with its natural check in the diminution of supply . . . We advise that all Acts of Parliament which profess to regulate, or restrict, the modes of fishing pursued in the open sea be repealed; and that unrestricted freedom of fishing be permitted hereafter.'

By 1999 the fish was almost gone, with only about one twentieth of the numbers in Victorian times left. At last Parliament woke up. Stringent rules were introduced on how many days' fishing were to be allowed each month, and the size of nets was reduced. That was not enough and great sections of the North Sea were closed to fishing. All this has made cod and chips a luxury and may soon turn it into no more than a memory.

Such is the price paid by those who ignore ecology – and when did you last eat a wild turkey?

To a Toga Party, as a Goat

The Spy's Curls back Maddener as she sings, 'The Girl with Colitis Goes By' in their new cover version of a Beatles classic.

For dyslexics, life is like that much of the time, with an unfathomable lack of fit between how a word is written and how it sounds. English spelling (which makes dyslexics feel like ghoti out of ghoti, or fish out of water – the fish with 'gh' as in rough, 'o' as in women, 'ti' as in nation; and the water with the same letters but as in plough, cord and tin) makes no sense.

Italian is much more logical. The Dantean '*Nel mezzo del cammin di nostra vità mi ritrovai per una selva oscura. . .*' both sounds and looks better than 'In the midway of this our mortal life, I found myself in a gloomy wood . . .'

The forty distinct sounds in English can be spelled in over a thousand ways. Italian is more economical in its utterances, with just twenty-five sounds and a mere thirty-three ways to put them on paper. Italy has, as a result, only about half as many dyslexics per head as does England, and Italian children learn to read their logical language well before their anglophone cousins.

Many of today's students claim to be dyslexic; they cannot spell or write decent prose and the diagnosis is, for some, sufficient excuse. Many of those with disorders of reading manage to overcome them with the help of extra time in exams, so that a simple deficit in intelligence is not to blame. My own test is to ask them to play Scrabble: a job that those with even a mild form of the condition find almost impossible. Often dyslexics reverse words or letters, or jumble them up. As they wander through their inferno of linguistic error dyslexics can console themselves that the wild complexity of their native tongue is at least in part to blame. A-level Latin – a language even more logical than Italian – might help their condition (and would certainly improve their prose).

The vagaries of the English language are not entirely at fault. Biology must play a part, for the condition is three times more common in boys than in girls. Brain scans of dyslexics show that, whatever their native language, something has gone awry in central command. The problem exists in Italy but, because of the beautiful rationality of the language of Dante, those who have it often do not recognise their state and can spell perfectly well. However, they can be tracked down with special word-recognition tests. Both groups – Saxon and Latin, those aware and those not aware of their condition – had an identical and unusual pattern of brain activity. They made less use of part of its left side, the main site of the language centre, which when damaged by a stroke interferes with the ability to read and spell.

The brain responds to the sound of music as much as to that of words, but in a different place. Its musical talents are less one-sided than those of speech, with plenty of cells that light up on the right side when listening to a tune. After a stroke, indeed, some people who cannot speak the words of a song can sing it with no difficulty.

The Italians are euphonious in tune as much as in text, and Joe Green (Giuseppe Verdi when you spell it right) himself wrote a church piece from a text by Dante. The English tend to be more like the old lady of Tring; who, when somebody asked her to sing, replied, 'It is odd, but I cannot tell "God, Save the Weasel" from "Pop Goes the King".'

Tune deafness of that kind is real enough. Play a distorted version of a familiar song and some people cannot tell it from the original. The condition, like dyslexia, runs in families but nobody has yet found out whether the brains of the tuneless work in a different way when they listen to music, or even whether Italians are, because of their musical heritage, less aware of the condition than are the English.

Can dyslexics sing and, if so, how do their brains light up? Perhaps they could learn to warble rather than write their exam answers (or one might teach the periodic table to the refrain of 'Pop Goes the Weasel'). Their brains might even be tested with a Spice Girls hit, except that nobody, whatever their nationality, could tell whether the tune had been distorted or not.

Bar Codes and Bytes

In 1999 a Nobel Prize went to the man who found out how proteins are sent to the correct places within cells. The work is rather technical, but several journalists came up with the same metaphor to explain it: each protein has a bar-code label attached, rather like that on a piece of airline luggage that allows it to be shuffled through the system until it reaches the right airport.

Such labels have become universal (and the bar-code metaphor is pretty tired, for the famous DNA fingerprint is often described in the same way). But what about bar codes themselves? Where do they come from? By an odd coincidence, just fifty years before the Prize, they burst on an unsuspecting world. On 20 October 1949 a US Patent entitled 'Classifying Apparatus and Method' was filed. It described 'article classification . . . through the medium of identifying patterns' and was the first attempt at a universal tag. The system involved four white lines on a dark background. It was based on Morse, with the dots and dashes extended to make broad or narrow lines, and a light-sensitive tube to read the reflection of a bright beam shone on to the label.

The first machines had a disconcerting habit of bursting into flames and it took twenty-five years before lasers and computers were cheap enough to allow the idea to be used in shops. The first commercial scan was done in 1974, at the Marsh Supermarket in Troy, Ohio, on a packet of Wrigley's chewing gum. The event was celebrated on its twenty-fifth anniversary in what may be the most boring exhibition ever (it had the very gum in pride of place) at the Smithsonian Institution in Washington.

Now five billion bar codes are scanned every day. Each has an internal logic. The numbers are split into two halves of six digits each. The first number is zero, the next five are the shop's own code (which allows it to sell 99,999 different items) and the remainder belong to the manufacturer. The system is rather complicated and is easily corrupted by dirt, a tear or a dishonest shopper armed with a pen.

As a result, the cryptographers of gum have come up with some safety features. The last of the twelve numbers in a bar code is a check on the others, an internal system of proof-reading that multiplies a chosen few of the earlier figures in the sequence together and subtracts the total from ten. The remainder left after the sum is done acts as a test, for it must fit the last number recorded in the bar code. If it does not, the label has been misread and the purchase is rejected. One time in ten the system fails as the check digit is, by coincidence, the same as that on the label, but such internal arithmetic remains a considerable help against accident or fraud.

The idea began in the supermarket but has spread much further. Cyclic redundancy code, as the system is called, is universal: every message sent from computer to computer has a block of information much longer than a bar code at the end that can be tested for accuracy with an internal calculation agreed by both sender and receiver. If it agrees, the message is accepted; if not, it has been corrupted and is thrown out.

And that is where the bar-code metaphor begins to take on a life of its own, for it has an uncanny similarity to the internal checks within each cell's Nobel-winning transport machinery. When proteins are made, an internal proof-reading mechanism – just like that on a frozen chicken – comes into play; every molecule is tagged with a hanger-on called ubiquitin. When it reaches its target, the check-in desk to the cell itself, the package is scanned by a piece of molecular machinery. If the tag fits – it matches the message made by an undamaged protein – the product is accepted, the tag is stripped off and the newly delivered molecule does its job. If it fails, the package is rejected and destroyed. Cells have a system of message verification that works like that of a supermarket. When it goes wrong, terrible things (cancer included) can happen.

No American shop has a bar code whose number includes 666, for a hundred million fundamentalists would refuse to buy it. The Book of Revelation prophesies that at the end of time we will all have to bear Satan's mark to be allowed to buy or sell – and what better than a bar code (perhaps based on one's personal DNA stripes) on the fore-head? True believers talk of a secret plan to tattoo an anti-fraud tag on to each of us. Then, if tattoo and credit card do not match, the heavies will be called in.

All nonsense, needless to say. However, St John's New Testament vision of Apocalypse was written on the Greek island of Patmos – and the Greek for the trolley that took his bar-coded baggage to the check-in is *metaphoros*: an object that helps one move forward. Just a coincidence, of course.

Fewer Bangs for the Buck

Homoeopathy's belief that weaker solutions make stronger medicine is rubbish; but what about medicine itself? We are all attuned to the idea that doctors with ever more powerful new drugs will soon keep us in full bloom until we drop at the age of ninety, but are we being realistic? Medicine has done a good job, for life expectancy has doubled over the past couple of centuries – homoeopaths, eat your hearts out! – but the story of its great advance is neither pure nor simple.

High-tech care is expensive. For health it might, as a result, seem obvious that more must mean better; and that rule works well enough on the grand scale. The unfortunate Sierra Leoneans can expect to live only to their thirties, while the Japanese make it to eighty. Global differences in expenditure are huge, with the richest countries spending hundreds of times as much per head on healthcare than do the poorest. The USA, home of the highest-tech treatments of all, comes top, with around $5000 per person per year; a quarter more than anywhere else apart from Switzerland.

It comes, then, as rather a surprise to learn that the United States is twenty-seventh in the world in terms of life expectancy, just one place ahead of Cuba (which spends about a twenty-fifth as much in dollar terms for each citizen).

Even in Europe, big bucks do not always bring good health. Britain ranks rather low in the spending stakes, with fewer than two hundred doctors for every hundred thousand people. Our life expectancy is seventy-seven years and several months. The Germans have twice as many physicians per head as we do – and for that great investment they live no longer than do the inhabitants of these rainy islands. In fact once national expenditure gets to about half the level of that in the United States the fit between what a nation spends on health and its citizens' length of life is weak. Above the threshold level a sort of financial homoeopathy rules: cash in makes no difference to health out. Africa itself has a better fit of life expectancy with national expenditure on education, particularly of women, than on hospitals and medicine.

Even so, in the lowest parts of the scale (and it is lengthy indeed, with Somalia blessed with only four doctors per hundred thousand) traditional economics – and medicine – works well. Worldwide, one death in three comes from infectious disease. Most could be prevented with some moderate expenditure on public health and, with the exception of AIDS, almost every one is, in principle, curable if caught early enough. Premature death from stroke, cancer and heart disease is often seen as a problem of the developed world – but plenty of fifty-year-olds in non-industrialised countries also suffer from such illnesses but cannot afford the medicines they need. For the poorest on Earth, more cash would mean much better health.

In most places the gaps in cash between people – and peoples – are closing. The average income of the poor has doubled since the 1970s and their health has, as a result, improved. An average child born at the turn of the millennium can expect to live eight years longer than one born in 1970. Much headway has been made among the very poorest, which means that much of the Third World is climbing the ladder to spending at Western levels.

The West itself has plenty of room for improvement. In the rotting slums of the Bronx male life expectancy is close to that of the Third World, while just a few hundred yards away, on Manhattan itself, are some of the finest hospitals on Earth. Britain, too, has great inequalities in health, for social and financial reasons alone. Everywhere in the richer parts of the world the urge to excel has meant that technology has been pushed to its expensive limit while simple – and cheap – preventive care has, in comparison, been neglected. High tech means high prices, with the USA spending a sixth of its gross domestic product on health, and Europe not very far behind.

The practice of medicine has nothing to learn from homoeopathy, the absurd idea that, for drugs, less means more. Even so, those who pay the bills should question the opposing belief that more must, of its very nature, always mean more. The internet price of a homoeopathic dose of Natrum Muriate (salt, for the uninitiated) is, as far as it can be calculated, five pounds a gram, ten thousand times the price in Tesco. Health economists, please note.

Potemkin Pillars and Replica Teeth

The London Borough of Greenwich is famous for astronomy, with Flamsteed, Harris of chronometer fame and much more. All this is familiar, but Greenwich – like many other parts of London – contains some unexpected scientific and technical treasures. Brunel's Thames Tunnel, the first to be dug under a river using a tunnelling shield, is still the only Tube line to have Doric columns and a moribund Victorian shopping arcade next to the track (it took so long to build that it became known as the Great Bore). The borough has plans for a modern technical marvel to match: a gigantic support bra built out of Kevlar in which to suspend that other Victorian icon of transport, the *Cutty Sark*, to ease the strain on her sagging curves.

Greenwich's greatest building is the Old Royal Naval College, which trained Britain's sailors for a century and a half. It is grand indeed, but (like many British icons) was built on the cheap. The 'marble' pillars in the chapel are, in the main, painted papier mâché. On its portico is a great tableau of the apotheosis of Nelson, with a ten-foot figure of our national hero at the centre. Many people assume it to be marble, but in fact it was made, in 1813, of a synthetic material whose secret was then lost for more than a hundred years.

'Chemical stone' was invented in the eighteenth century by Mrs Eleanor Coade, who left her father's Devon farm and set up in business in London. She developed – and nobody knows quite how – a complex ceramic which was years ahead of its time: a mixture of clay, flint, sand, soda-lime and silica glass softened with linseed oil and fired at more than a thousand degrees for four days. Her factory was close to where the South Bank headquarters of what was the Greater London Council (abolished by Mrs Thatcher) stands today. At first the company made clay images, and its artisans spat on their work as they modelled it. Commissioned to produce a head of George III, Mrs Coade insisted on the use instead of rosewater sprayed from a silver syringe. So impressed was the King by her good manners that her commercial future was assured. The ingredients were mixed, pressed into moulds and then fired.

Coade stone was remarkable stuff. It was tough, had crisp edges, resisted frost – and, over the years, has dealt with London's toxic air far better than most natural materials could manage. Great architects – Robert Adam included – loved the new medium, and many rivals tried to copy it. All failed, and the secret of how to make plastic stone died not long after its inventor. The last pieces were manufactured in 1843.

Almost by accident, Eleanor Coade had hit on a formula for a synthetic composite whose properties far transcended those of its individual constituents. The idea is ancient enough – the Babylonians knew how to make bricks with straw – but put her on the cusp of the modern world.

The Age of Metals peaked in about 1945 and since then technology has gone backwards, into a second era of composites. Brick, paper and china are examples of the genre, but their modern equivalents are far more sophisticated than before. They do remarkable things. The Airbus fuselage has composites where the 747 has metal alloy and is much lighter as a result. In jet engines, too, a high-temperature ceramic reinforced with silicon-carbide whiskers does not crack and can be used in places where the toughest metal would melt. Gears and cogs made from such materials no longer need to be machined at great expense but can be moulded. One clever composite has embedded in its structure a lubricant that is released as it wears away, to eliminate the need for oil. Others have built-in sensors able to respond when they exceed safe levels of temperature or stress.

In the world of medicine surgeons have made artificial bone stronger than its natural equivalent. Dental crowns used to break after too solid a bite. The problem was crack propagation: once started, the fracture spread through the porcelain, with its regular arrangement of atoms, at the speed of sound and a great chunk fell off. The problem was solved with crowns made of resin filled with glass-like particles that are more similar to natural teeth, which have protein fibres able to stop fractures from spreading.

Twenty years ago, and after much trial and error, Mrs Coade's magical material was recreated, and one can now buy expensive copies of the great urns and fountains designed by Robert Adam and by lesser mortals when the new material was first made. Sad to say, they are dark grey, so that any attempt to put teeth back into the Greater London Council will not succeed.

Squeezing Blood out of a Cow

B-plus: the story of my life; and of my blood group. B Rhesus positive is the aristocratic mix borne by just one Briton in fifteen. Blood groups have always attracted bad science: one massive best-seller on diet and blood type recommends that Bs should avoid rhubarb, and most of the population of Japan is convinced that people of that blood group are blessed with cheery and creative personalities (they are, of course, quite right).

Below the froth lies a great reservoir of solid research. It started long ago. In 1666 Robert Boyle, of gas-law fame, wondered whether a mastiff's blood transfused into a bloodhound might reduce its ability to smell (and his friend Samuel Pepys speculated that if he moved blood from a Quaker to an archbishop the latter would change his beliefs). A year later, on the other side of the Channel, a Monsieur Mauroy was given calf's blood in the hope that an infusion of the animal's mild nature would calm down a 'phrensy, occasioned by a disgrace he had received in some Amours' (syphilis, in other words).

He died, of course, for his immune system recognised the cow's blood as foreign. Even a transfusion from a random neighbour had a good chance of killing him, for those who give must match those who receive. There are thirty or so separate blood group systems – some familiar, like ABO and Rhesus, and some less so – and we have little idea why they are there. Rhesus is the hardest to understand, for a marriage between a positive man and a negative woman may lead to the mother mounting an attack on her unborn Rhesus-positive child, with disastrous consequences. Evolution might be expected to do away with one variant or the other – but it has not.

ABO is also hard to comprehend. There are dozens of associations with disease, for people with O resist malaria well but do worse than others at cholera, while the As suffer more from cancer (and we Bs are, some say, at higher risk of diarrhoea) – but a difference in disease susceptibility does not explain why all the diversity is there in the first place.

Doctors prefer 'How?' questions to 'Why?' queries like these and their sanguine attitude has paid off, for the battle against blood groups is well advanced, on several fronts. Eight million people have a transfusion each year, and there is a worldwide shortage of the crucial fluid. Blood replacement is a hot topic. The search for an alternative began with Sir Christopher Wren, who suggested that beer would work. It doesn't; but the military began to pour money into the problem at the time of the Vietnam War and the research has begun to bring dividends. Now we can extract the crucial oxygen carrier – haemoglobin – from the cells that hold it. Then its molecular units are linked into a chain to add stability to the protein, and the naked substance injected straight into the patient. This avoids the problem of blood-group matching altogether.

The method has real advantages. Haemoglobin can be stored on the shelf for years, is safe from viral contamination and causes no immune response. Although it does not last long in the body and can damage the kidneys as it breaks down, it is already much used in South Africa (where there is a shortage of safe blood because of the AIDS crisis).

The most remarkable fact – with a nod to the memory of M. Mauroy – is that, in one version of the treatment, the haemoglobin used comes not from men but from cows (and Jehovah's Witnesses, who reject human blood, really like the idea). Human haemoglobin genes have now been inserted into bacteria and a version of the protein, subtly modified to improve its ability to stay in one piece as it floats free in the circulation, may soon be available in large quantities.

If bacterial haemoglobin seems a step too far, the Americans have made an entirely synthetic replacement for the vital substance. It is related to the banned refrigerator fluid Freon and has, in its latest version, ten carbons arranged in a chain, with a series of fluorines attached. The compound soaks up lots of oxygen and releases it on demand. Already such chemicals have been tried on about fifty thousand patients, with some success.

There remain, though, certain difficulties. Will those who have a transfusion of O blood (accepted by everybody, whatever their own group) become, as the Japanese believe, more dependable? And is there a danger of turning bovine after a shot of haemoglobin or chilly with a dose of the artificial stuff? I rather hope I never need to find out.

Waxing Warm and Cold

Global warming is a real concern, and vast sums have been spent on wind farms, wave-driven power stations and the rest in an attempt to avoid it. It would be far easier, and much more efficient, to save energy and cut down on greenhouse gases by insulating a few houses. Even better, they could be built to save energy. The ecologically sound house is, thanks to the laws of chemistry, nearer than many people think.

Lightweight modern buildings heat up and cool down at a much greater rate than do solid old structures. An ancient stone or brick building has a large thermal mass and can, once at the right level, keep a constant temperature without too much heating and cooling. Today's houses lack inertia, the ability to hold a desired temperature – warm or cold – once it has been reached. A flimsy modern bungalow could, in principle, be made just as ecologically friendly as the Tower of London with an enormous rock in the kitchen; but the object would be inconvenient at best (and rocks are expensive).

Some architects have tried to make such an energy reservoir with water, which takes up heat in the warm and releases it as the temperature drops. However, the volume of water needed makes it impractical for general use (unless it is kept in an indoor swimming pool). Water, though, hints at another ecologically friendly approach to heat control. Like all liquids, it has a complex relationship between temperature and the amount of heat stored. A garden pool covered with ice warms up as the temperature rises during the day;

but it pauses as the ice melts because to liquefy water once it is in its solid form itself takes energy. At night the pond cools down again, but again there is a halt in the rate of temperature change as the liquid shifts back to solid.

It takes a lot of effort to make or break the ice, for to melt ice to water, with no increase in its recorded temperature, uses as much energy as to heat the same volume of ice-cold water to the level of a central-heating radiator. To keep a house at the traditional British point of just above freezing one could, as a result, store the heat not in a large bath of water but in a small fraction of that volume of melting ice.

In these decadent times people ask for a warmer place to live. We need something that shifts from solid to liquid at a higher temperature than does water. Such a material would also be useful in hot weather, for it would cool the air at no cost.

German chemists have come up with a paraffin-like substance that moves from waxy to oily at the appropriate point, soaking up energy as it does. As the day gets warmer, the excess energy goes into melting the wax rather than heating the air; and the heat is released at night as the liquid slowly solidifies. The material is held within tiny balls in a liquid matrix. It is powerful stuff, for when mixed with plaster and applied to a wall as a coat one inch thick it has the same heat-storage capacity as does a brick wall one foot deep. It works in winter too, as rooms do not get too hot when the heating goes on and freeze when it switches off. A change in the number of carbons will shift the point of transition, so that each country, or household, could be provided with a version set at the temperature it prefers (and makers of refrigerators are keen on the cooler versions).

This material is already in use on a housing estate in

Germany and – with the help of triple-glazed windows and solar heating – has cut heating costs by nine-tenths, to about twenty-five pounds a year. Soon, promises the company that makes it, only one litre of oil will be needed for each square metre of floor space, which will give a bill of just ten pence a week. However, the idea is a bit *Vorsprung durch Technik* for us. We were, after all, the nation that stuck to open coal fires – the least effective form of heating known to man – long after most of Europe had moved to stoves, some insulated with brick to hold the heat for longer.

Why not continue to do it the British way – make long johns out of the stuff, or flat caps, or mufflers? Why not thermal socks, or those nice rugs that grannies used to put over their laps as they peered into a smoky fire? In fact science could help us make a giant stride backwards – and what Briton could ask for more?

Accentuate the Negative

Email spam is a real annoyance, but most companies and universities have filters that do pretty well at stopping it. In the early days they simply picked up offending words (Viagra, penis, teens and all the others never used in polite conversation) and blocked the message. The approach worked for a while, but the junk soon crept back, with some odd additions. The nasty stuff was surrounded by strange phrases such as 'plangent chondroitin epicanthic rugosity', or by blocks of familiar but subtly enhanced prose: 'It was the best of times, the worst of times, increase penis length click here.'

Well, Dickens never wrote that (although, given his own sexual eccentricities, he may well have thought about it), so why clog up the world's in-boxes with such stuff?

It has to do with the way in which the spam-filter hunts its prey. Even the most innocent email must now and again include the word 'Viagra' or 'penis' or mention teens, and it can be annoying to have it deleted just because the censor noticed one forbidden term. Instead it scans the whole text and assigns a spam score to each word. Add up the scores in relation to the length of the message by using a certain statistical rule and if it exceeds the threshold, out it goes.

Bayesian statistics – upon which the rule is based – are named after the English cleric who invented them. They combine separate bits of information to make a better guess about what might happen next. They are much used in genetics. Imagine that someone is at risk of having a child with an inherited disease. A standard exam question goes: 'A woman has a one-in-four chance that her first child will be affected. In fact it is healthy; what is the chance of illness for her second?'

The answer might seem to be one in four again, on the casino analogy, for an even number on the roulette wheel does not change the chance of 2, 4, 6 or 8 over 1, 3, 5 or 7 on the next spin. But that is too simple. If the lady has ten healthy children in a row it becomes almost certain that she does not carry the hidden gene and that her eleventh will be healthy. Every infant adds a bit of information and even just one normal child reduces the risk for the next. The same is true for email: 'cheap', no problem, 'Viagra' alone OK, 'buy on-line' just acceptable; but the three together within a couple of sentences means doom.

The artful spammers use other tricks to imitate nature. A good camouflage strategy for a tasty insect is to look like a random sample of the wider background. The best place to hide a leaf is in a tree and a crude copy on an oak is safer than an almost perfect one on a daffodil. It is all very Bayesian; for a predator, if the first ten leaves are just leaves, then so – probably – is the next, so why not stop looking?

Spammers, as a result, conceal their diseased mail among information that looks healthy; random words not in an opponent's dictionary, or lumps of innocuous text, so that he lowers his guard and lets the junk through. It is a matter of hiding the tasty target in verbiage and jamming enemy statistics with a mass of irrelevant information (which explains the 'plangent chondroitin epicanthic rugosity').

Even so, they face a problem. If the magic key to virility is concealed within too much text, the recipient himself may not spot it. There, too, those who poison the world's electronic wells have begun to imitate nature. Mimicry, like camouflage, involves two players, the viewer and the viewed. Because animal vision varies so much from species to species, what is persuasive to one creature is crude to another. My favourite example is the bee orchid: a flower that looks slightly, but not much, like an insect although the presence of leaves, stems and the rest makes it clear to us that it can be sniffed without too much danger of being stung.

Not to the male bee. Its simple mind is so fixed on sex that the flower is the entomological equivalent of Viagra. It sees only the image of a potential mate and copulates as hard as it can, picking up pollen as it goes.

The latest trick is crude but effective and the smartest statistical package can do nothing about it. It changes the typeface of part of the message to white to match the screen's background. You – the viewer – see, like a randy bee, only the joyous news: 'Instant orgasm guaranteed!' while the filter in its plodding way reads the digital code all the way through: Instant (long section of the Book of Isaiah), orgasm (a great deal of Shakespeare) guaranteed. Its censor nods, and the message lands in your in-box.

Why not just give up and click on the link?

Hunting in Groups

All Londoners have the uneasy feeling that beady and critical eyes are tracking their every move. Plenty of them belong to surveillance cameras but most are attached to pigeons. Throw down some bread – no, please don't; it is against the law, leave it to the old ladies with plastic bags – and within seconds, out of nowhere, dozens of the birds appear.

How do so many find their crust so soon? Do they all watch everyone all the time (particularly people with bulging bags)? If so, it must keep the birds busy and might interfere with their other hobbies.

In fact the pigeons' surveillance system is much more efficient. They keep watch not on humans, but on each other. Once one individual spots food and flies towards it, his friends (or rivals) follow at once. As a result, a dispersed web of spies turns within seconds into a frenzied mob of consumers. Advertisers use the same logic as they scatter their wares before the public: persuade a few movers and shakers to eat a certain brand of cereal and the rest of us will change our breakfast routines at once.

Many animals spy on their fellows and, as a result, soon pick up a set of shared habits that help them cope with life. Rats have sharp teeth, sharp eyes and sharp noses. Like pigeons, they follow their comrades to food, but when the animals are faced with a new and unfamiliar item they smell the breath of their neighbours to see if they have tried it. If they have, the food is probably safe and the hungry rat tucks in. If, on the other hand, the rat next door looks under the weather – perhaps because he has eaten poison – his crafty fellows pick up the deadly scent and avoid the bait thereafter.

Group feeding is fun and so is group sex. Female antelope, like rats, are fond of perfume; they sniff the soil for the heady aroma of a pregnant animal's urine to find where their sisters have done their mating and hang around the sacred spot on the grounds that that is where the boys are. Female guppies are just as shameless: when they see a particular male hard at it they offer themselves to him with no fuss about courtship – for he must be pretty hot if he has already attracted a partner.

The cynical biologist can easily change their preferences. For most of the time the females go for bright-coloured and aggressive suitors. But arrange matters so that they see some dull and earnest male copulating away while his flashy fellows are sidelined and they soon develop a strange interest in piscine nerds (which is good news for scientists, at least). Guppies prove that nothing succeeds like success, and that success feeds on itself. A successful male gets lots of females, his rivals copy his behaviour and in time a whole new sexual culture sets in.

Such trends can lead to geographical changes – local quirks in birdsong, for example – in which the statements of sexual interest vary from place to place. Whales and some songbirds have dialects of this kind, and paternity tests show that, in a particular place, males with the wrong accent have fewer young than do those who speak in the approved way. The barriers held in the mind can be hard to penetrate, for in some birds at least the edges of each dialect are also areas of genetic change.

Could this apply to us? Well, maybe; after all, to sing 'I

Belong to Glasgow' has more of a chance of attracting a mate in Sauchiehall Street than in Soho, and the opposite is perhaps true for 'Maybe It's Because I'm a Londoner'. Any visit to a pub in Edinburgh shows that the effects of local mate preference are obvious enough, but whatever the differences in courtship behaviour, the Scots and the English are not separated by impregnable sexual barricades, and the genes show no real change at Carlisle or Berwick.

In contrast, marriage records over the centuries do reveal a strong impediment to sex across the language divide between the English- and Welsh-speaking parts of Pembrokeshire, a barrier that has been there since Flemish weavers arrived nine hundred years ago. An old piece of research hints that there may be a matching shift in blood groups but the evidence is thin and needs redoing with modern technology. On the European scale, too, language barriers at the old frontiers are often matched by genetic change.

I once gave a talk in Belfast about such matters. At the end a host of hands went up. Almost all belonged to people asking the same question: 'Are there any genetic differences between the Catholics and the Protestants?'

Science has not yet answered that question and when it does will do nothing to solve their frontier problems.

Vultures, Cultures and Criticism

Critics are the occupational hazard of every author, for they are drawn to literature as vultures to carrion. To the herbivores who produce their raw material they are about as welcome as a circle of hungry hyenas must be to a wildebeest who is feeling a bit down in the mouth; but in fact, like scavengers, they play a useful (albeit repellent) part in the intellectual food chain.

My own vitals do not remain ungnawed and the annoying thing is that sometimes the vultures get it right. I once attempted to write an update of *The Origin of Species* and, yes, it was a mistake to call Erasmus Darwin Charles' greatgrandfather (one too many generations there) and when I talked of ancient penguins around the Mediterranean I meant polar bears. *Mea culpa*, and I put them right in the reprint. There may be too many jokes in my homage to the great man (what use is a book without jokes?) and – no question, I admit it – the book was not as good as *The Origin*. How could it be?

But the odd thing is that so much of the criticism focuses on something not discussed in my book at all. I missed out, so it says, the most important aspect of modern evolution, the part that Darwin would have been most excited by, the subject that sets the intellectual world abuzz. The crucial topic is the attempt to explain human behaviour and society in terms of natural selection. To omit it shows cowardice, ignorance and narrow-mindedness. Why put in so much about fossils and miss out the nuclear family? Doesn't everyone see that society is coded in the DNA; that, like the wings of the melanic moth, politics is just a way of increasing reproductive success?

I missed the topic out for two simple reasons: because there is no mention of it in *The Origin* and because the attempt to explain the modern world in terms of the sex life of the Stone Age is the biggest load of hogwash ever foisted on to an unsuspecting public. There is a severe and unavoidable danger of Arts Faculty science – wonderful theories, facts not required – when using the lessons of animal behaviour to understand human affairs.

Of course, some of our behaviour does descend from our ancestors, which is why solitary confinement is so painful. Had we evolved from an unsociable primate like an orang-utan we would no doubt punish criminals by making them go to Booker Prize dinners instead. And, I admit, the new science of the evolution of human behaviour has made some breakthroughs, for now we know that mothers love their children and that older men fancy younger women.

It is good to have such things confirmed with rigorous research, but once one moves on from the blindingly obvious the whole thing turns into a parlour game, with prizes for those who can make up the most plausible story (no evidence required). Acne – reduces sexual allure of those too young to care for children; blushing – signal from female of readiness to mate; crime – male display of readiness to take risks; and so on to zoophilia (use your imagination).

What has this got to do with science? It degrades the whole field of sociobiology, which was started by Darwin himself in his discussion of sterile insects and how they help their kin. The subject has become one of the most exciting parts of genetics, with ingenious theories and experiments to match; and, like any science, lots of unanswered questions.

Compare that with the palace of strawless bricks that

makes up most of evolutionary psychology. The origin of society must be, one assumes, quite a complicated problem. I am astonished by how simple the explanations are and how uniquely low are the standards of proof demanded. Perhaps its supporters have been wandering in a Freudian fog for so long that even pseudo-science looks attractive. A vague acquaintance with Darwinism and a boundless self-confidence are the only qualifications needed to become an expert.

Part of what we are is, needless to say, in the DNA, but to say that the genes are in control has the same intellectual content as saying that I went to work this morning because my legs dragged me to the bus. The problem arises from a failure to understand the methods of science. A theory must be disprovable for it to have any value, but for the enthusiasts everything we do must have evolved because, well, otherwise it wouldn't be there, would it? For my part, I prefer to put in more about fossils.

Flanders and Swann saw the problem of the endless argument from nature. In their song 'The Reluctant Cannibal' a father, exasperated at his son's refusal to eat a tasty piece of flesh, comes out with some biological logic: eating people is right, he says, because otherwise God would not have made us of meat. Critics, no doubt, say just the same to their children.

In the Shadow of Science

The arts are always in the penumbra of science, for the ability to appreciate them is set by the physical limitations of the aesthete within the skull, otherwise known as the brain. Shadows themselves have a long history in both fields of endeavour, and they go even further, for they illuminate some of the murkier aspects of human belief.

A full Moon looks flat, benign and almost supernatural. True devotees saw it as a heavenly body, a jewel in the sky that in its holy perfection stood in stark contrast to the flawed Earth below. In 1609 Galileo with his telescope put an end to that comforting thought, for he saw long streaks of black thrown by mountain peaks in the lunar evening which proved, to general astonishment, that the Moon was not some kind of divine crystal but a new and rugged world not much different from our own. He even worked out that its mountains were no higher than those on his native planet.

Over the next few centuries astronomers made hundreds of striking images of our satellite viewed as it waxed or waned. The shadows cast at lunar sunset allowed them to build up precise three-dimensional maps of a body hundreds of thousands of kilometres away. Today, in an even more mysterious insight into the universe, quite small galaxies seem to cast huge shadows, far greater than expected, as light from more distant places streams past them; perhaps because they, too, have a hidden and massive structure – dark matter – whose existence, like that of the mountains of the moon, can be seen only when they interrupt the beams of a distant star.

Shadows cast light on to grey matter as well. Conspiracy theorists, like medieval theologians, are happy to dismiss the evidence of their eyes if it clashes with their own creed. The *Apollo* moon landings were, they say, faked on a giant stage in California. One favourite piece of evidence is that, in the famous photograph of the first flag on the Moon, the shadow of Buzz Aldrin is much longer than that of Neil Armstrong. The paranoid claim the existence of two spotlights; but physicists point to a minor case of the Galileo effect – for just behind Armstrong the Moon's surface rises, shortening the dark patch made as he blocks the sun.

Shadows can be confusing things. Art uses a sort of virtual geometry that could not work in the real world but is accepted as reality when hung on a wall. Paintings fool those who see them into the impression that a few daubs on a flat surface are a three-dimensional scene. As a result, they test the limits of what the brain is willing to recognise as truth.

By the time of Galileo painters could deal with distance – perspective – pretty well, but many still struggled with light and shade. They drew shadows that fall in different directions within the same scene, with some even going round corners. Wrong though they were, the sophisticated eye of the modern aesthete scarcely notices, although their occasional clumsy failures in perspective are painfully obvious. Apart from that, the brain seems not to care much about shadows, perhaps because they do not say much about what is going on in the real world.

Such things are holes in light; and, like any hole, are best defined by what they lack – depth, colour or internal structure. Their shape depends on how bright the light is, where it comes from and on the surface upon which they fall. Computer manipulations of size, colour, graininess, shape and so on find that our perceptions insist on only one thing: a shadow must be darker than its surroundings to be recognised for what it is. Nothing else seems to make much difference.

Plenty of illusions turn on the brain's dislike of that insubstantial world. So powerful is its aversion that the grey matter prefers to interpret whatever it sees to be a solid body when it gets a chance. Take two identical objects – two paperbacks, say – and up-end them next to each other in sunlight so that they cast identical shadows. Draw a line around one – and at once it looks darker and more real than the other, and now resembles a paper cut-out rather than a mere absence of light. Even a few squiggles of ink inside its limits give it a more permanent air.

Shadows do add something to a painting, for if they are removed from a standing figure it can appear to float in the air. Vampires do not have this problem, for they lack such constant companions to anchor them to the ground. In religious works certain figures show their holy state in the same way, impossible though that is in the real world. At least Galileo had the common sense not to bring that up when he faced the Inquisition.

Evolution According to Ovid

One of the off-Broadway hits of 2002 was *Metamorphoses: a Play*, a series of vignettes based on Ovid's fables. At eighty dollars a seat the Midas touch was called for (and his was the first tale) but it was worth every cent. Most of the stories concern a shift of identity: live girl to dead metal, narcissistic shepherd to elegant flower and drowned sailor to bird soaring in its Halcyon days. In homage to its watery theme the action took place in and around a swimming pool and was so energetic that front-row tickets came with an intriguing note: 'Customers may get wet. Towels are provided.'

Ovid's myths depend on the perennial magic of seeing one form of life change into another. It is a trick much used by novelists ('As Gregor Samsa awoke one morning from uneasy dreams he found himself transformed in his bed into a gigantic insect') but is also of interest to science. How does a caterpillar reinvent itself as a butterfly, or a tadpole as a toad?

We have started to find out. The truth turns – in Ovidian style – on mortality foretold. If you are a fast reader, your body has made about two million cells since your eyes strayed to this page. All but a couple of thousand of them will soon die, in a planned execution called apoptosis (after the Greek for leaf-fall). The jury is made up of genes – reaper, grim and toll, among many others – who could themselves be characters in Greek tragedy. The executioner is an enzyme called caspase. He cuts his victim into ribbons, starting at the nucleus.

In myth and molecular biology alike death is a prelude to

new life, for metamorphosis – from caterpillar to butterfly or from man to bird – calls for the planned destruction of obsolete cells. We ourselves have webbed hands and feet before we are born, but cell suicide removes those sheets of skin before birth (Cygnus, in the Ovidian version, reversed the process and turned into a swan). Certain chemicals confer unwanted immortality by blocking the machinery. They give, for example, a toad with a tadpole's tail and might some day make a mouse with webbed feet.

The aged couple Baucis and Philemon, whose sole request to the gods was to be allowed to expire at the same moment, could now be helped by science, for the links (and they are complicated indeed) in the lethal chain have been untangled. They involve an ordered set of hormones and other chemicals. Although the details vary from creature to creature these grim jurists all have a segment called the death domain, which – as Ovid would be delighted to learn – is identical through much of the living world. As a result, the proteins that kill unwanted cells and transform men's bodies also work in worms.

Cells commit suicide for other reasons: when they find themselves in the wrong place (skin cells survive only when surrounded by their fellows and fall on their swords should they wander into the body itself), when the DNA is damaged or when – as in the immune system after an infection – their job is done. A refusal to turn up the toes when ordered to can lead to cancer; while an undue desire to end it all causes degenerative illnesses such as Alzheimer's disease. Programmed cell death, in its Gothic complexity, is of much interest to medicine although few drugs have yet been targeted at its machinery.

Lots of Ovidian characters (Baucis and Philemon

included) turn into trees. They look forward to a literal apoptosis in the autumn of their lives, but one of the bathing deities had a different fate. Phaeton, son of the Sun god (and brother of Cygnus), was cast in the play as a spoiled schoolboy floating on a sunbed. Awarded quality time with his estranged father, he treks to Phoebus Apollo's cave and is offered, in Manhattan style, anything he wants. And – what else would you expect from a lad desperate to copy his dad? – he asks for the keys of the car.

Alarmed, Apollo warns him of the dangers of a solar chariot in inexpert hands ('Spare the whip! Take the middle way!') but Phaeton sets off in great confidence across the sky, only to swerve off the track. So wild is his course that he sterilises Libya and boils the Euphrates. At last, in an attempt to save the world from universal fire, the gods unite to destroy the vehicle, and – in more of a metaphor than a metamorphosis – Phaeton meets his programmed but miserable end. A pity, really, that he did not listen to their advice in the first place.

Staying Upright on Burns Night

I do not own a kilt; nor am I often seen in diamond-patterned harlequin pantaloons in a saucy red, white and blue, but I once came across both those sartorial rare birds at roost in the same habitat. The occasion – a Burns supper in a snowbound Scottish castle. The observation – a difference in the preening behaviour of those clad in either garment. And the science – an insight into how the mind gets confused, on Burns Night most of all.

Such events are often ghastly and the after-dinner entertainment gives new meaning to Wodehouse's 'shifty, hangdog look which announces that an Englishman is about to speak French'. Change 'Englishman' to 'Scot' and 'speak French' to 'indulge in country dancing' and you get the flavour. Before the reeling can begin the sporran must be combed, the dirk polished and the kilt worn with the gravity it deserves. That means *straight*: hem parallel to the floor, with the squares of the tartan arranged with edges vertical and horizontal, and no Sassenach concession to the quirks of the human frame.

All this took a lot of adjustment among the snows before things looked right and the comedy could commence. The prancing pantaloonists, on the other hand, bent their flexible limbs at once, with minimal need to tweak the nether man, which appeared smart at all times.

Their elegance is the product of an illusion, both in the mind and below the waist. A tilted kilt grates on the eye more than does a mistuned pantaloon because it is far easier to see a departure from the general pattern in an

arrangement of vertical or horizontal lines than in a slop-
ing one. As a result, a flaw in a criss-cross diamond fabric
is far harder to spot than is one in a square and stolid
tartan.

It all turns on the brain's ability to interpret the outside
world. It was discovered long ago that particular cells in a
certain part of its cortex respond to lines that slope at an
angle, while others go for the vertical or the horizontal.
That second class of sensitive cells is much better at the job
than are the relatives who concentrate on the oblique; and
imaging shows that the brain is forced to work harder when
presented with a vertical grid than one leaning over at forty-
five degrees. The emblem of St Andrew (who was,
according to unreliable sources, crucified on an X-shaped
cross) has, modern science proves, less of an identity than
does that of St George.

Our preference for the upright makes us much more sensitive to deviations in a vertical than in an oblique pattern. A
single sloped dash in a page of verticals hence stands out like
a sore thumb, while a vertical dash surrounded by slanted fellows is harder to spot. At a Scotland–England rugby match a
Scots supporter waving his flag at the foreigners' end will, as
a result, stand out more than a red cross flaunted among a sea
of saltires. For the more brutal among the English fans, the
whole visual system follows the rules: seeing the Caledonian
intruder, moving the gaze towards him and thumping the
unwelcome visitor – each action done better when the eye is
presented with an oblique flaw in a vertical background than
when it faces the opposite challenge.

All this makes sense in the world of the senses. The horizon is for most of the time – well, 'horizontal' is the word,
for trees grow up and apples fall down. Rather fewer things

come to our notice at an oblique angle. A whole host of illusions is born of that simple fact.

Such quirks of the brain lead to a whole host of illusions, optical and otherwise. I subscribe to a common one: that true happiness is to be found only in France. I have a modest house there (and you should see my face when I try to speak the language) and have just added an extension, a simple brick box. The job was badly done and the room was not square, a fault that became obvious when large earthenware tiles were laid out. At once the eye complained at the obvious flaws at the edges, where the walls intruded at an angle into the grid. The local artisan, using tradition rather than science, had the answer – lay the flooring at forty-five degrees to the edges of the room, to give a diamond pattern which, to those insensitive cells that specialise in the oblique, is quite acceptable. The builders, of course, were English.

Beer Gives You Energy

My physics teacher in my Liverpool schooldays was a Mancunian in the worst sense of the word. Some effete southern snob had already introduced us to his subject but, he discovered, had led us into terrible error. The noble son of the Irwell made his fury clear: 'It is not JOOL's law, you cretins, it is JOWL's! Born and died in Salford! A brewer; haven't you heard of JOWL's Stone Ale? What do you measure energy in [metre stick poised at the ready]? Yes, in bloody JOWLs! JOOLs are for poofs!'

Strange what makes a memorable lesson. I have vivid memories of the stick (excruciating), the pronunciation (eccentric) and the ale (extinct), but I had to remind myself what Joule's law actually is (it relates the amount of heat given off by a current to the resistance of its electrical circuit).

James Prescott Joule was a synthesiser, and of much more than beer. He brought economics to the world of physics, to their mutual benefit, when he realised that the universe had a common currency, called energy. The Joule is immune to inflation (conservation of energy: what goes in comes out) and lives in a free market, for it can be converted into heat, work, motion, electrical power and more. On his Swiss honeymoon Joule found that the temperature of the bottom of a waterfall was higher than at the top as the energy of the falling water was transformed into heat. Electricity follows the same rule (which led him to invent arc welding), as does gas as it expands (which in time gave us the refrigerator).

His notion has moved into biology. An animal is just a

machine to transform capital from one form to another; the energy held in food into bones, muscles or bad temper. Different creatures live at different rates. It would be more expensive to keep an elephant-sized mass of mice as pets than to keep an elephant, for to be a rodent takes many more Joules per ounce. That is why mice are such pesky pests; they need a constant supply of high-energy fuel to keep them going.

Neither mice nor elephants would do well in the competitive world of nineteenth- (let alone twenty-first-) century Manchester. A stationary steam engine in the Joule brewery could manage around 25 per cent efficiency, while a modern gas turbine (which spins directly in the flame and uses any waste heat to run a second, steam-based turbine) can extract about 80 per cent of the energy in its fuel. A mouse does no better than 3–4 per cent, and an elephant even less.

All this has an effect on the economics of nature. It controls the number of steps in the food chain. Most places have no more than five steps between the bottom and the top; from the plant soaking up sunshine to the tiger that feasts on flesh. The production line is so inefficient that it cannot afford to get too long. Business itself follows similar rules. A product is made from raw materials, packaged, delivered and sold. Low-grade stuff in fields (barley) is converted into a high-grade product in pubs (beer). Each step calls on the great god Joule to provide energy. All companies are worried about the length of their supply chains; the more links between factory and shelf, the more the waste. Cutting out the middleman pushes up efficiency – and profits.

The parallel between ecology and economics goes further. The best predictor of food-chain length in a lake is not energy, but size, and the bigger the body of water, the more the steps. A trout in a small pond feeds further down the chain than does one in Lake Erie. Size means efficiency and big lakes, even hungry ones, have more species (just as big companies like Tesco can stock more products than a corner shop, however well the latter controls waste).

Not all the similarities between biology and business are comforting. It seems obvious that a complicated place like a rainforest would be better buffered against disaster than is a simpler one like a tundra – but the evidence is far from clear. Large and complex communities with lots of inter-linked food chains are liable to undergo unexpected and catastrophic shifts. The Sahara was a grassy plain for five thousand years, but turned for no obvious reason to desert in a couple of hundred. A rich and varied ecosystem, in nature or in commerce, is just as much at risk of random disaster as is a poor and simple one – as investors in Enron soon noticed.

Joule's brewery was taken over by Bass Charrington in 1970. Bass is now just one of fifty or more beers sold by a giant consortium called Interbrew, and holders of Joule Brewery stock pin their hopes on what has become little more than waste paper. They can always turn to drink for consolation.

Paranoia Makes Us What We Are

I once reviewed Richard Dawkins' book *The Ancestor's Tale* in the medical journal the *Lancet*. It is a good read, as one would expect, but I noted its several mentions of President Bush, which – in spite of the remarks often made about his supposed resemblance to various primates – sit somewhat oddly in a tome on evolution.

In return I received an email from a professor at Johns Hopkins University (a major American centre for biological research). He did not like the piece. 'Little Stevie,' he began (not a good start), 'What I found was the grumbling of yet another toady prig who gets his jollies from Bush-bashing. Where did you cower during the London Blitz? Thank God the Atlantic separates us.'

I was, I admit, no hero in the Blitz as I had omitted to be born in time, but stung by his comments I fired off – as one does – a brief and rude reply stating my intention to stay on this side of the ocean until the Bush gang had been dismissed from power. The distinguished professor's response was a bit of a surprise: 'Have you thought of taking psychiatric advice for your delusional, paranoid disorder? Just in case you might change your mind about a visit, I'm going to have our Homeland Security put you on the suspicious persons list.'

False denunciations to the secret police are a reminder of a time and a place when paranoia was justified, but medical advice from a physician at Johns Hopkins is not to be set aside lightly. I know no psychiatrists, but can turn instead to people in my own trade for assistance. They find an unexpected link between paranoia, the feeling of persecution often associated with schizophrenia, and the evolutionary ascent, if that is what it is, from monkey to academic.

Those two creatures, even at Johns Hopkins, look much the same (as Gilbert and Sullivan put it, 'Darwinian man, though well behaved, at best is but a monkey shaved'). What makes us different is in the brain. More and more genes have turned up which differentiate our own grey matter from that of our unshaven relatives.

Humans have unique mutations in lengths of DNA associated with, for example, brain size or the ability to speak. One gene helps determine the amount of nervous tissue and has evolved at enormous speed since the split from chimps. Any damage returns a child's brain to the volume of that of its ancestor, with drastic results. Sometimes the effect is more subtle, for in the human lineage certain genes have been doubled up compared with their primate relatives, another tactic to increase the pace of life within our skulls.

One piece of DNA codes for a brain enzyme responsible for recycling a chemical messenger called glutamate. The gene is present in double copy in all apes, but no monkeys. As a result, the flow of information through ape – and human – brains speeds up in comparison with all other mammals, which must been important in the intellectual leap forward that took place twenty million years ago when our apish ancestors broke away from the monkey line.

It seems that at least some cases of schizophrenia are involved with changes in just the same gene. A study in Russia (plenty to worry about there!) shows that patients with paranoia had a more active form of the crucial enzyme than did unaffected people. Perhaps their feelings of persecution come from a failure to control their mental activity within normal human limits, so that the natural caution we all share has turned into a pathological fear of being followed, or lied about to the secret police.

Schizophrenia, a frightful illness, affects around one person in a hundred. It involves much more than changes in a few genes. Even so, the famous glutamate pathway may be important to all of us, schizophrenic or not. A variant in a second gene that controls the activity of the brain messenger is borne by less than half the population and appears to protect against the disease. (I do not know which form I carry, although it would be easy to find out.) Some drugs used to treat schizophrenia do their job by interfering with the flow of the chemical and – early days as these are – scientists are hard at work on new compounds that might do the same, but better.

Whether I will turn to the professors of Johns Hopkins for diagnosis (and whether that university is happy to have its name attached to crude threats against foreign academics) I prefer not to discuss here, in case the information is used against me. Or am I being paranoid?

The Battle of the Dinosaurs

Deep in South Kensington lurks a monster; a vast petrified beast that has struck terror into generations of young people. It is Diplodocus – Dippy the Dinosaur, who dominates the main hall of the Natural History Museum. At eighty-five feet and ten tons he might seem a fearsome creature, but in fact Dippy was rather a nice, Bloomsbury kind of dinosaur, snacking on vegetables and fond – had the tea plant evolved in time – of a few gallons of Earl Grey to wash them down.

The museum also houses a high-tech and heartless relative: an animatronic *Tyrannosaurus rex* who roars and gnashes on demand. That alarming animal is best visualised as an Armani-plated predator who tore great hunks out of any vegetarian foolish enough to get too close.

In 2003 the conflict moved from the plains of ancient Wyoming, where those creatures once found their home, to the intellectual foothills of London, for a plan was put forward to merge two of its universities – Imperial College (science-based and close to Dippy) and my own stamping ground, University College London – into a forty-thousand-strong conglomerate, a great sauropod of pedagogy. The idea sprang fully formed from the fertile brows of Messrs Richard Sykes and Derek Roberts, the heads of each college.

It caused panic among the grazing giants of Gower Street (at South Kensington the animatronics must have been out of order, for our twin institution remained frozen in glacial passivity). The case 'for' turned on cash and diversity (degrees in structural engineering with structuralism?), that 'against' on identity, history and more. The merger would help Imperial to shake off its image as an Institute for Depressed Chemists, brown coats abulge with Bunsen burners, while UCL might gain its partner's email address, ic.ac.uk, which sounds like a cat being sick.

Such issues loom large for those who sip Earl Grey in the common rooms of Bloomsbury, but may mean less to the rest of the populace. But they raise a wider question – why gigantism? Is there a natural trend to large size, and is Big always Bad?

The truth is far from clear. Animals cut off, like academics, from the real world can either grow or shrink. The Seychelles have a giant tortoise but a frog the size of an ant; and that rather confused beast the pygmy mammoth flourished on islands off Siberia just when Ireland was infested by gigantic elks. Dippy's own ancestors went from six to forty feet in just a few million years, but his successors shrank. Some say the giants grew to defend themselves against competitors (the logic of the present merger), but others blame the low quality of what they were taking in (food in their case, students in ours, but no comments on A-level standards, please).

Dippy himself was a Dinosaur of Very Little Brain. The contents of his skull in relation to body size were less than a fifth the size of the equivalent in his tyrannical relative. So vast, indeed, was the animal that it needed a second nerve centre in the seat of its pants. *T. rex* was a much sharper operator because he was small and quick on his feet (a fact lost upon the financial merger maniacs of the 1990s) – and it is worth remembering that the dinosaurs saved themselves by growing wings and flying away (we call them birds nowadays).

In education the story is much the same: the small specimens soar while the leviathans lumber. Harvard has about the same number of students as UCL (albeit with a lot more cash) while the largest university in Europe is Rome's Università La Sapienza, which – sapient as it might be – is not, with its hundred and fifty thousand undergraduates, a major intellectual player.

Even so, the merger plans ground on. New Gargantua would need to rebrand (although we who speak from the pants of the beast still like the label used against our secular ancestors in 'the godless institution in Gower Street'). A title that accepts both UCL's tea-bibbing history and IC's megalomania might be the 'Grandiose, Imperialistic and Nebulous Giant Establishment, the Roberts–Bentham Institute of South Kensington, United in Torment' – or GINGER BISCUIT for short (the substitution of C for K comes from Latin, the language of academic discourse).

The new identity provides both a coat of arms (Gold, two scaly lezards combatant vert) and a motto, *Iunctus in Tormentum*. Others may prefer Consignia College, or Enron University. Most of us, though, just wished we could get back to Diplodocus' day, in the Jurassic, seventy million years before *Tyrannosaurus rex* appeared on Earth. Fortunately, the whole crazed scheme collapsed under its own contradictions and – for the time being – the vegetarians have inherited the Earth and dinosaurs still graze in Gower Street.

A Taxonomist's Noel

Science, like sex, is often fun but rarely amusing. Christmas is just the opposite. Scientists who try to be comic, like clowns at parties, tend to plough sadly on, regardless. The low point comes when they name plants and animals. The need for millions of Latin terms, and a free hand in so doing, has led to great families of names, some droll, others much less so, from the orchid *Aa* to the jellyfish *Zyzzyzus*. Some, like *Scrotum humanus* (once thought to be the intimate remnants of a giant but in fact the leg joints of a dinosaur) are unwittingly witty, but most make too much of an effort. Even so, with the help of *Sanctacaris* (a primitive animal, long extinct, best translated as Santa Claws) one can, with a mussel-bound *Abra cadabra*, conjure up a taxonomist's dream Christmas.

Who to invite to the feast? *Mamma* the snail and *Gramma* the fish will insist on coming, whatever the salamander *Oedipus complex* might say. For more attractive company one could try the molluscs *Amanda, Daphne, Doris, Fiona, Julia, Mathilda, Melanie, Patricia* and *Sallya*. If snails are not enough, invite the fish called *Clara, Liza* and *Rita*, not to mention *Diana* (*Camilla*, alas, is a kind of fly). For consorts, any old *Ptomaspis, Dikaspis* or *Ariaspis* (fish fossils: try dropping the specific name of the viper, *aspis*) will do.

And what about dinner? One could be safe and try *Cannabis* (a tasty bird) followed by *Ambrosia* (not creamed rice, but knapweed). As an alternative, while slaving at the kitchen *Formica* your ant might whip up a dish of molluscan *Exotica* or a *Box* of fish. If not well cooked, this brings the danger of making the guests feel waspish (*Verae peculya*) or even of the onset of a bout of *Emesis*, *Sepsis*, *Dialysis* or *Townselitus* (all of them insects). In the worst of all worlds, the meal might be followed by an attack of *Dyaria* (a moth named in honour of the eminent Dr Dyar), which could force the unfortunate company to face the *Enema pan* or the *Colon forceps* (both beetles) before they fall off their perch and turn *Beliops*.

Oops (a spider; the name also given by mistake to a beetle), what a *Neardisaster* ('no point around the star'; the shape of this starfish's mouth) that would be! And then there is the dreadful stuff on television. During the *Hiatus* (the fly) before the Queen's speech it is bound to be something like *Batman* (a fish), *Draculo* (another one), *Godzillius* (a crab), *Cinderella* (no flies on her), the spiders *Draculoides* (*bramstokeri* is the scary one) and *Apopyllus now*; or the three *Muscatheres* (just three species of these bee flies). Later there will be reruns of fossilised humour – it is *Montypythonoides* (a dead snake) again.

Then we have to listen to *Papa* (the bird's) endless *Saga* about cricket, which always ends in *Anticlimax* with a fossil snail. Perhaps we can bug him with a quick snog (*Ochisme*, anyone; *Polychisme*, *Dolichisme*, *Marichisme*, *Nanichisme*, *Florichisme*, *Peggichisme*?) beneath the mistletoe before flying (*Iyaiyai*, *Ohmyia omya*!) upstairs in the hope of a pitch-pine (*Pinus rigida)* bedstead and a quick opossum or *Philander*. Then it's time for a toast (*Veni vidivici* if you're an extinct parrot) to the wasps – *Heerz lukenatcha*, old boy; *Heerz tooya*, too! – and to perch with *Banjos* before bursting into song. *Ia, Ia io* has a batty refrain, as do *La cucaracha* and *La paloma* (not cockroaches or doves but moths) and the great horsefly *balzaphire*. A piece by the musical bee *Mozartella beethoveni*, or the chiggers *Trombicula doremi* and *T. fasola* for seashell *Chorus* accompanied by snails on *Tuba* might follow. As a *Bonus*, the limpet could sing a tree *Aria*, although that raises the danger of such *Cacophonia* that everyone clams up.

At Christmas, the scarab *Euphoria* in its leafhopper *Nirvana* soon gives way to *Hades*, the *Notoreas* butterflies from hell. *Lucifer*, *Satan* and *Mephisto* (more fish), the shark *Gollum* and *Caligula* (a moth) make their metaphorical appearance. Under these circumstances, who can blame the deranged biologist as he exclaims: '*Bloodiella*!, *Pisoneu*! you parasitic wasps! *Bugeranus*, you wattled cranes, and *Ba humbugi*, you snails from the island of Mba!, you can't be the cactus *Cereus*! *Ytu brutus*, I intend to beetle off; I'm going to *Dissup irae*!' (a very small insect).

He conquers his *Agra phobia* (see below), puts up his snail *Umbrella* and leaps into his weevil (or *Car*) – be it *Amercedes* (another weevil), a Beetle (*Agra vation*) or a fishy old *Sierra* – and drives with a bee fly of unorthodox tastes to the *Villa sodom* for a curry, pausing on the way to pick up a can of Indian beans from his *Lablab lablab* . . .

Enough, enough! I invoke the Sanity Clause (thank you, Groucho) in my book contract.

Index